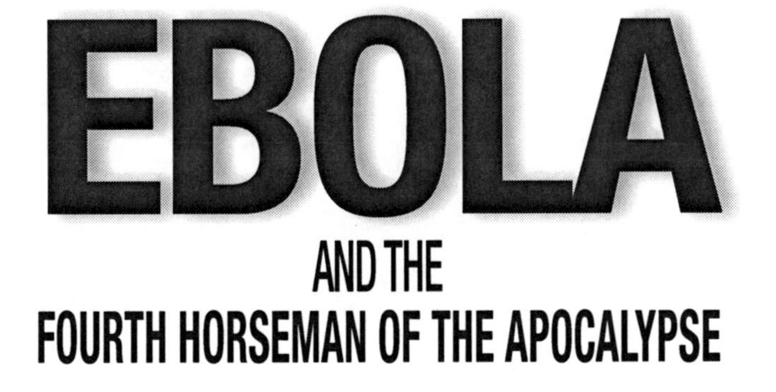

EBOLA

AND THE
FOURTH HORSEMAN OF THE APOCALYPSE

Defender
Crane, MO

Ebola and the Fourth Horseman of the Apocalypse

Defender

Crane, MO 65633

Printed in the United States of America.

ISBN-10: 0990497445

ISBN-13: 978-0990497448

A CIP catalog record of this book is available from the Library of Congress.

Cover illustration and design by Daniel Wright: www.createdwright.com.

All Scripture quotations are from the Holy Bible, Authorized King James Version.

This book is dedicated first and foremost to the Bride of Christ, those precious men and women, teens and children who have placed their faith in a Risen Savior and look for His returning with glad anticipation. And to my loving husband, Derek, who embodies Christ and His love to me every day of our married life together.

Foreword from Tom Horn

First of all, I'm not a scientist. Before all this happened in West Africa, I could barely spell Ebola; ok, I could, but I certainly had no idea how powerful, how infectious, and how merciless this thing called Ebola Hemorrhagic Fever could be.

I'm just a preacher, you see, but it doesn't take a theologian—nor does it take a biologist like Sharon K. Gilbert to see that something wicked is on its way. Sharon does understand the science, and you will find this book inspirational, informative, and a little bit terrifying. But Ebola is not our real enemy. Scientific ignorance is understandable, in fact, it's the norm for most of us. However, ignorance of Biblical prophecy is killing the church. That's the worst epidemic of all.

Churches across America are being led—or rather misled—by men and women who are not much better than confectioners, because they spend every Sunday passing out candy to a congregation that's so full of sweets already that they're passing out from diabetes! It's time pastors and teachers step up to the plate and get it right. You can't preach a partial gospel any more than a husband and wife can be partially married. You either preach it all or forget it—you either make a woman your bride, or you don't. End of story.

So why then do so many in today's candy-store pulpits leave out the prophetic books of the Bible? Our scriptures begin with prophecy, and they end with it. In Genesis, God promises a Redeemer to Adam and Eve—a prophecy that was later fulfilled when Jesus Christ became a human and died for our sins, becoming the 'New Adam'. It ends with prophecy, because the book of Revelation is the guidebook, our map to the future—a future that, I'm telling you, is looming right over the horizon, folks. We Christians need to get it right—right now. Preachers need to preach it right, *right now*—because time is running out.

West Africa is swimming in blood, and that blood is testifying that prophecy is coming true. This book will help you to understand the science of Ebola, but it is also a primer on how the Ebola crisis is a foreshadowing of the coming Horsemen. Read it, then give it to your loved ones to read, because, dear friends, it is coming. You and I and all the world had better get ready, because the Horsemen are about to ride.

Tom Horn,

October, 2014

From the Author

We're all scared, aren't we? I know many with whom I talk about the subject of Ebola and the current outbreak, particularly as how it may affect our neighborhoods, our towns, say they are terrified. But every day in life brings its burdens, its unforeseen traps, tricks, and terrors. A day that begins with laughter over breakfast can end in tears by midnight.

As Christians, we have a promise, made to us by a God who never changes, never lies, never deceives, never tricks. He is the same today as He was yesterday, and He will be the same tomorrow and forever. He loves us enough to send His Son to die for us, even while we were sinners—just so you and I can have life, and hope, and a promise for eternal life in His presence. Paul says this in his epistle to the Romans:

> Who shall separate us from the love of Christ? Shall tribulation, or distress, or persecution, or famine, or nakedness, or peril, or sword? As it is written, For thy sake we are killed all the day long; we are accounted as sheep for the slaughter. Nay, in all these things **we are more than conquerors through him that loved us.** For I am persuaded, that **neither death, nor life, nor angels, nor principalities, nor powers, nor things present, nor things to come, nor height, nor depth, nor any other creature, shall be able to separate us from the love of God, which is in Christ Jesus our Lord.**
>
> —ROMANS 8:35–39 KJV (Emphasis added)

The content is this book may overwhelm you, may cause you to wonder if God truly is here—is truly looking still upon the world to which He once came as a suffering servant—if He still intends to keep His promises. But I tell you that He has not changed, He is still with us just as He promised when Jesus told His followers that He would never leave them without a Comforter. God is here in the person of His Holy Spirit, and He knows about every virus, every infection, every victim, every nurse, every doctor, every report, every rumor, and every letter of every word that has been or will be written by a human—right down to the bits and bytes, ones and zeroes that inhabit the Internet.

He knows your fears—He knows of your sleepless nights, your worries about where our world is headed and how it will affect you, your children, and your grandchildren. Jesus Christ loves our families far more than we can ever love them, for He is our Creator, our

Bridegroom, and our King. He is our Savior, and He is coming back to set all things right. But before that glorious day, there will be times of stress and trial, but do not let your hearts be troubled, dear brothers and sisters. Jesus holds us within the palm of His hand, and He will never forsake His promises. Nothing can separate us from the love of God. Nothing. So as you read this book, keep that thought uppermost in your mind. Memorize the scripture from Romans. Write it on a slip of paper and tape it to your fridge, and store it up in your heart.

Trouble is on the way, but our King stands ready to meet it, and He has given each of us His protection. Jesus Christ is coming back to this Earth soon, and woe to the Troublemakers who have tried to hurt those whom Christ loves. Jesus said it best:

> Let not your heart be troubled: ye believe in God, believe also in me. In my Father's house are many mansions: if it were not so, I would have told you. I go to prepare a place for you. And if I go and prepare a place for you, I will come again, and receive you unto myself; that where I am, there ye may be also.
>
> —John 14:1–3 KJV

To this, we the Bride must say with joy in our hearts, "Even so, come, Lord Jesus."

And so it begins....

And when he had opened the fourth seal, I heard the voice of the fourth beast say, Come and see. And I looked, and behold a pale horse: and his name that sat on him was Death, and Hell followed with him. And power was given unto them over the fourth part of the earth, to kill with sword, and with hunger, and with death, and with the beasts of the earth.

—Revelation 6:7–8

Death is a hungry Serpent that bites in the night. The hapless victim is completely unaware of the microscopic, coiled presence as its poisonous form nestles into a warm epithelial, endothelial, fibroblast, or nerve cell—and it begins to make itself cozy and comfortable. Or better yet, the Serpent may bite into a macrophage or dendritic cell; those cells that live and travel throughout the human body, acting as stalwart sentinels, the cellular equivalent of watchers on the wall, who report back if an attacker is spotted on the horizon or near the gates.

But this attacker is something I call the Blood Serpent, the *Filovirus* known as Ebola, and it knows how to cripple the body's own SOS signals to send the victim's body into a cascade of excruciating pain, fevers, and internal bleeding—turning the terrified victim into a zombified biological bomb. And when this bomb goes off, the victim is dead, and all those around him have been bitten.

It is early October as I write this book. This is my favorite time of year. The air begins to chill, and the trees show off with panoramic displays of colors no human artist can ever truly match. As I peer out my window, I can see our cul-de-sac neighborhood swathed in layers of crisp, brown leaves that rise up into miniature spirals whenever the breeze blows. It's cool now, so our windows are open. I can hear the little girl next door laughing as she plows through a pile of leaves, her black Labrador playmate close on her heels. She's two years old, hair the color of beach sand, and she knows no fear.

She should. Because an ill wind is blowing across the Earth, and it carries sickness, death, and panic. This morning I watched in awe as a lunar eclipse turned our pale moon to blood. Tonight at dusk, faithful Jews around the world will begin a joyous, weeklong observance of Sukkot, the Feast of Tabernacles, as the Christian and Jewish worlds mark the second of a *tetrad* of four, blood moons. There've been numerous articles, YouTube videos, and several very well-written books about this marvelous clue to God's timeline. Joel 2:21 warns us that:

> The sun shall be turned into darkness, and the moon into blood, before the great and the terrible day of the LORD come.

This warning is repeated, word for word, in Acts 2:20:

> The sun shall be turned into darkness, and the moon into blood, before that great and notable day of the Lord come:

And in the book of Revelation, the Apostle John sees the ultimate fulfilment of this prophecy:

> And I beheld when he had opened the sixth seal, and, lo, there was a great earthquake; and the sun became black as sackcloth of hair, and the moon became as blood…

The blood moons are harbingers—signals that the time of the end is drawing near. The rising up of the Bride is perhaps one breath away; and then, the four horsemen, who comprise the first four seals found in the book of Revelation, will begin their terrifying, destructive rides.

Or have they already begun? Some contend that the rides commenced the very moment that Jesus Christ, the risen Lamb of God, arrived at His Father's throne; that it's His triumphant return to Heaven as the Risen Lamb that John sees in Revelation 5:6:

> And I beheld, and, lo, in the midst of the throne and of the four beasts, and in the midst of the elders, stood a Lamb as it had been slain, having seven horns and seven eyes, which are the seven Spirits of God sent forth into all the earth.

Jesus Christ is the Lamb, who is the only one deemed worthy to open a great 'book' or scroll—essentially, a legal document that transfers the ownership of Earth and those within it to the care of the One who created them. Adam and Eve, you see had been given authority over the Earth, but they chose to disobey God and place their trust in the lies of the *Nachash*[1]—that is, the Serpent—and by doing so, they relinquished their authority over the Earth to the Snake.

But not for much longer. The triumphant Lamb will very soon return to Earth and reclaim what Adam threw away, and whether Christ began to open that scroll then, or is only now about to crack open that final seal—the outcome is the same: Earth is about to enter a phase of testing and judgment that will make all previous wars, catastrophes, and plagues seem like a picnic.

And the hallmark sign of these troubling times, so to speak, are these moons of blood. Since the first blood moon shone upon the Earth during Passover in the spring of this year, the world has learned of a tragic epidemic taking place in West Africa. It is an epidemic of blood. And as of this date (October 9th), over 3400 people have died. The CDC and WHO (Centers of Disease Control and World Health Organization respectively) estimate that the reported numbers may be inaccurate, that the real numbers could be three times greater.

On October 1st, we learned that the United States has confirmed a case of Ebola Virus Disease (EVD) here—in Dallas. The victim arrived from Liberia after telling airport authorities there that he had not been near or touched an Ebola patient. He had apparently lied. Thomas Eric Duncan had, in fact, played the hero and carried an infected pregnant woman—a dying woman—to a local hospital. Perhaps, after that, he panicked and decided to leave Liberia on the next flight. Perhaps, Duncan reasoned that the US could provide far better healthcare than the shanty care centers in most of poverty-stricken Liberia. Perhaps, reports of the survival of a nurse and a doctor, both of whom had worked for Samaritan's Purse, contracted Ebola and returned to the US for treatment was playing inside Duncan's mind—and he rushed to Dallas, where he had family, including a former girlfriend, Louise Troh, whom he was reportedly hoping to marry.

Whatever the reason for his sudden departure from West Africa to Texas, Duncan's story is one that is very likely to be repeated—perhaps has already been played out again and again, and we simply don't yet know it. If you lived in Liberia and had the money to leave, wouldn't you?

This book is about our current crisis. Though most people in the world do not yet know the Blood Serpent is slithering into their neighborhoods, they soon will—although by then, it may be too late. So, what do we do? How can we prepare? What is Ebola, and why do I compare its emergence with the book of Revelation and blood moons?

As the angel said to John again and again in the book of Revelation: "Come and see!"

Now the serpent was more subtle
than any beast of the field.
—GENESIS 3:1

It is thought that viruses similar to Ebola have lurked within the shadows of our jungles and caves for millennia. Some theorize that the earliest known outbreak of a hemorrhagic fever virus may have been the Black Plague, because contemporary diaries and medical descriptions of the disease include signs and symptoms indicative of a hemorrhagic event. The Great Plague of Athens may also have been, in truth, an HFV (Hemorrhagic Fever Virus) event.

Following is a translation of Thucydides' notes on the 430 B.C. outbreak:

> As a rule, however, there was no ostensible cause; but people in good health were all of a sudden attacked by violent heats in the head, and **redness and inflammation in the eyes**, the inward parts, such as the throat or tongue, **becoming bloody** and emitting an unnatural and fetid breath. These symptoms were followed **by sneezing and hoarseness, after which the pain soon reached the chest, and produced a hard cough**. When it fixed in the stomach, it upset it; and **discharges of bile of every kind named by physicians ensued**, accompanied by very great distress. In most cases also an **ineffectual retching** followed, producing **violent spasms**, which in some cases ceased soon after, in others much later. Externally the body was not very hot to the touch, nor pale in its appearance, but **reddish, livid, and breaking out into small pustules and ulcers**. But internally it burned so that the **patient could not bear to have on him clothing or linen even of the very lightest description; or indeed to be otherwise than stark naked**. What they would have liked best would have been to throw themselves into cold water; as indeed was done by some of the neglected sick, who plunged into the rain-tanks in their agonies of **unquenchable thirst**; though it made no difference whether they drank little or much. Besides this, the miserable feeling of not being able to rest or sleep never ceased to torment them. (Translation by R. Crawley, in M.

I. Finley's *The Viking Portable Greek Historians,* pp. 274–75)[2]
[Emphasis added]

I've added bold type to several of the symptoms, including red eyes, bloody cough, diarrhea, pain to the touch, ripping off clothing due to being 'hot' internally, and bloody pustules on the skin. These are all hallmarks of HFV. Ebola patients in all African outbreaks have often stripped off their clothing, becoming violent and trying to escape their beds and enclosures as if the virus were driving them to go outdoors and spread the contagion.

Titus Lucretius Carus wrote a poem about the Athens Plague including symptoms that are all too familiar to those suffering an HFV epidemic:

If any then
Had 'scaped the doom of that destruction, yet
Him there awaited in the after days
A **wasting and a death from ulcers vile**
And **black discharges** of the belly, or else
Through **the clogged nostrils would there ooze** along
Much **fouled blood**, oft with an **aching head.**[3]

Ebola is but one of a larger family of hemorrhagic fever viruses (HFV) which include the filoviruses, arenaviruses, bunyaviruses, and flaviviruses. Arenaviruses include (year discovered in parenthesis) LCMV (1933), Junin (1958—Argentina), Machupo (1963—Bolivia),

Lassa (1969), Guanarito (1989—Venezuela), Sabia (1993—Brazil), Chapare (2004), and Lujo (2008).

The natural hosts for arenaviruses are usually rodents of some kind, which live near or inside victims' homes and urinate/defecate onto beds, furniture, corners, etc.—which allows the virus to become airborne whenever someone sweeps. Such an airborne virus can then enter human nostrils and gain access to the lungs, or the microscopic particles can land in food or water, which then gains the virus entry into the gastrointestinal system.

Bunyaviridae (bunyaviruses) include Hanta, Dugbe, Bunyamwera, Rift Valley, and Crimean-Congo. Bunyaviruses are transmitted by arachnids, primarily ticks. However, Hanta is a rodent-borne disease, which is most memorable for the epidemic that broke out in the American southwest in the 1990s. *Sin Nombre* (Spanish for 'nameless') was what the Mexican-Americans called the unknown killer that struck in the Four Corners region where Colorado, New Mexico, Arizona, and Utah meet. According to the CDC, the fatality rate during the *Sin Nombre* outbreak of Hanta was almost 67%. The deer mouse was determined to be the natural host, and urine/feces the means of infection. Spring cleaning had created airborne virus particles that infected nasal passages and lungs as women and children swept their homes.

Hanta Virus had received its name earlier in the century during its initial known outbreak in South Korea near the Hantan River (it is often customary to name an HFV for the nearest river or town). The South Korean event ran from 1950-1953 and initiated the modern battle against HFV. More than 3,000 troops, both American and

South Korean developed the disease, which had a ten percent mortality rate, considered low for HFV, which can run as high as ninety percent. Karl M. Johnson, an American virologist, partnered with South Korean virus hunter Ho-Wang Lee to ferret out the natural host for Hanta: the striped-field mouse.

Flavivirus family members (*genera*) are well known to most who follow the news. These include West Nile Virus, Dengue, Tick-borne Encephalitis, and Yellow Fever. Flavus is the Latin term for 'yellow', hence the family name. Victims are often jaundiced as a result of liver impairment, giving their skin a yellow cast. In 1793, Yellow Fever—sometimes called the American Plague—killed over nine percent of Philadelphia's population. Arthropods and arachnids carry flavivirus to humans, mosquitoes and ticks being the primary culprits.

The final family of HFVs is the Serpent. Why do I refer to this particular virus in such a manner? Because it looks like one. Morphology (or the 'shape') of the other virus families are round (like a small sphere or even a koosh ball), but Family *Filoviridae* are called ropes, shepherd crooks, worms, or snakes due to their unusual appearance. These have a linear portion that ends either in a curved hook-like appendage or a circle (sometimes called the 'Cheerio').

It is this shape that most impressed me while researching the current outbreak back in March of this year. I was preparing two talks for the Prophecy in the News (PITN) Pike's Peak Summit, a prophecy conference that included myself, my husband Derek P. Gilbert, and an entire smorgasbord of prophecy scholars from across the nation, headlined by Gary Stearman, LA Marzulli, Bill Salus, and many others. It was a privilege to be included in such a stellar list, so I wanted to bring something new and important to the 'table'.

Since my background is biology and healthcare, and since I'd been keeping an eye on what was then a mild outbreak of Ebola in Western Africa, I decided to learn as much as I could about the pathogen. At that time, Ebola Virus Disease (EVD), was unknown in West Africa, having previously been isolated to Sudan, Uganda, and Zaire, but you can appreciate how this disease could easily spread from one war-torn country to another.

Because the outbreak at that time seemed to be gaining steam, I decided to write a presentation on the topic of emerging diseases and the Fourth Horsemen of the Apocalypse. This required digging into Revelation 6 and the original language used to describe the horrific, futuristic vision given to John by our Savior, Jesus Christ.

Want to know more? Come and see…

And when he had opened the fourth seal, I heard the voice of the fourth beast say, Come and see. And I looked, and behold a pale horse: and his name that sat on him was Death, and Hell followed with him. And power was given unto them over the fourth part of the earth, to kill with sword, and with hunger, and with death, and with the beasts of the earth.

—Revelation 6:7-8 KJV

Before we discuss the meaning of the Rider called Death and his pale horse, let's examine the first three horse/rider combinations that are revealed to John by Jesus Christ in Revelation 6.

As I mentioned previously, there are a few prophecy scholars who contend that Jesus Christ began to unseal this legal document (the scroll) upon His arrival at the throne of His Father in the 1st century, just after His ascension into Heaven. We must remember that God's throne is not restricted to our space/time limitations, so the throne has a perspective to all time periods. God created time for us. He is not bound by time; He exists outside of time. John experienced these visions and wrote them down in or about the year 96 A.D., but his arrival in Heaven took him outside of his time/space constraints, so the visions could have been contemporary to the 1st century or taking place in a far, distant future.

The majority of prophecy scholars contend that John was writing of a future unsealing (relative to his own time period), and that the Riders and all that followed in their wake would commence near the 'time of the end' or in 'the last days'. The first three chapters contain letters dictated to seven churches in Asia Minor (modern day Turkey) by the Lord Himself to John. Afterward, John then in Chapter 4, we read:

> After this I looked, and, behold, **a door was opened in heaven: and the first voice which I heard was as it were of a trumpet talking with me; which said, Come up hither**, and I will shew thee things which must be hereafter.

How marvelous! This passage is a picture of what Paul told the church in Corinth:

> Behold, I shew you a mystery; We shall not all sleep, but we shall all be changed,
>
> In a moment, in the twinkling of an eye, at the last trump: for the trumpet shall sound, and the dead shall be raised incorruptible, and we shall be changed.
>
> —1 Corinthians 15:51–52

And again, in his letter to the church at Thessalonica, Paul writes these words to comfort them:

> For this we say unto you by the word of the Lord, that we which are alive and remain unto the coming of the Lord shall not prevent them which are asleep.
>
> For the Lord himself shall descend from heaven with a shout, with the voice of the archangel, and with the trump of God: and the dead in Christ shall rise first:
>
> Then we which are alive and remain shall be caught up together with them in the clouds, to meet the Lord in the air: and so shall we ever be with the Lord.
>
> Wherefore comfort one another with these words.
>
> —1 THESSALONIANS 4:15–18

Paul is describing an event that we often refer to as the 'Catching Away' or the 'Rapture' of the Church, the Bride of Christ. No, the term 'rapture' does not appear in the English. The Greek term translated as 'caught up together' is *harpazo*, but the Latin equivalent in the Vulgate translation is *rapiemur*, from which modern English speakers derive 'rapture'. Both refer to suddenly being 'snatched up'.[4]

Most prophecy scholars believe that John's vision of the seals and all that follow will not begin until *after* the Rapture of the Church occurs. I agree with this position, that the Bride of Christ—the Church—will rise to meet Him in the air before the commencement of the so-called 'Tribulation Period', which is just another name for the final week of Daniel's Prophecy of Seventy Weeks.[5] The Blood Moon[6] tetrad (four lunar eclipses that all occur during Jewish holy days over the space of two years) that began earlier this year may be a major sign that the Final Seven Years is about to begin and that Christ will soon return.

For the sake of simplicity, we will approach the unsealing events from the more commonly accepted perspective that this has not yet occurred but is imminent. It's best to quote the verses as we discuss them. In verse one of Revelation 6, we see Jesus Christ opening the first seal of a legal document, which only He is worthy to unseal (picture a scroll that is sealed at various points with wax and is rolled up, and the first seal is visible on the outside of the scroll that our Lord holds).

He breaks the first seal to open the document:

> And I saw when the Lamb opened one of the seals, and I heard, as it were the noise of thunder, one of the four beasts saying, Come and see. And I saw, and behold **a white horse**: and **he that sat on him had a bow**; and **a crown was given unto him**: and he went forth **conquering, and to conquer**.
>
> —REVELATION. 6:1–2

This first horse (Greek *hippos*) is white—in Greek *leukos*, means shining white, brilliant, and it most often refers to the brilliance of angels. It is the same color as that of Christ's horse described in Revelation 19:11, when he returns to rule the Earth at *the end* of the

Tribulation Period (aka the Seventieth Week of Daniel's Vision). At that time, Christ will return with His Bride and usher in one thousand years of peace and prosperity.

However, this first rider is an *imitation* of Christ. He carries a bow—a word with a double meaning. The bow can mean a weapon, or it can mean the bow (rainbow) that shines from God's throne. This double-meaning is intentional, for it reveals to us that the unsaved will not perceive this rider as a conqueror but as a ruler. The Rainbow coalition springs to mind as an example of a political faction which uses the Rainbow as a symbol of 'many peoples'.

The Greek word translated as 'bow' is *toxon*, a very curious word which literally means 'bow', but is derived from verb *tikto*, which means 'to travail, to bring forth, to give birth'. This picture of childbirth echoes a warning given by Jesus to his followers and to all of us; that TEOTWAWKI ('The End Of The World As We Know It') will commence after much travail.

For when His disciples asked what would be the signs for the coming of 'the end of the age', Christ replied:

> For many shall come in my name, saying, I am Christ; and shall deceive many. And ye shall hear of wars and rumors of wars: see that ye be not troubled: for all these things must come to pass, but the end is not yet. For nation shall rise against nation, and kingdom against kingdom: and there shall be famines, and pestilences, and earthquakes, in diverse places.
>
> All these are the beginning of **sorrows.**
>
> —MATTHEW 24:5–8 (Emphasis added)

Christ tells them that imposters would rise and try to fool the people, wars would break out, famines and pestilences, but all these would be just the beginning of the 'sorrows'. This word sorrows is *odin* in the original Greek, referring to 'intolerable anguish' or 'the pain of childbirth'.

Christ is telling us that the Tribulation Period will be preceded by birth pangs. If you're a woman who's given birth, or if you have witnessed childbirth, then you know the intense pain that precedes and accompanies this marvelous miracle. You may also know that sometimes birth pangs are felt weeks before a child is due to be born. Most often, these early pangs are called 'Braxton Hicks'[7] contractions. They feel like the real thing, but they are just a reminder that a child is on the way.

However, the arrival of the horse and the first rider bring *real* birth pangs. This rider and steed combination are harbingers of something about to be born—a New Age, when the enemy—that old Serpent—will set up a false kingdom and attempt to unseat Christ. Remember, that in Matthew 24, the disciples had asked Jesus to tell them when the 'end of the age' would occur. As I write these words, we are closing in on the final moments of our current age, and a new one is about to commence, but Christ will be its ruler, not Satan. Satan will try to preempt God's plan, but Christ's return will stop this seven-year lie in its unholy tracks.

Returning to Revelation 6 and the unsealing of the legal scroll, we find horse and rider number two:

> And there went out another horse that was red: and power was given to him that sat thereon to take peace from the earth, and that they should kill one another: and there was given unto him a great sword.
>
> —REVELATION 6:4

This horse is red. The Greek here is *pyrros*, literally meaning the color of fire. This horse blazes with flame, and its rider has been given power—that he should take peace from the earth.

Red is also the color of human blood, and it is the free flow of this precious life-giving force that will overflow the Earth during the Red Horseman's ride, for he is given a great sword, a *machaira*. This word refers to a 'large knife' used for cutting up flesh, or a small sword. This is a picture of a machete, a small blade that is straight or curved and is used even now in African and Middle Eastern nations to hack into dense jungle or to hack up an enemy.

In April of 1994, on the seventh day of that ominous month, a radio signal was heard across the African country of Rwanda; a short message that was the code phrase to start a genocidal massacre that would shock the world. The station was RTLMC (Radio Télévision Libre des Mille Collines) a French station which broadcast from July 8, 1993 until July 31, 1994. The name of the station is based upon the well-known description of Rwanda as 'the land of a thousand hills'.

The station played popular music which drew in younger listeners, but music was interspersed by vitriolic harangues against the Tutsi tribe members who lived and worked alongside the Hutu.

Tutsi people were called cockroaches, and the vile rhetoric broadcast on the station advocated crushing them like bugs.

When on April 6th, Rwanda's President Juvenal Habyarimana was killed in a plane crash, Tutsis were instantly blamed, and the station called for all-out war, using what is perceived by some historians as a code phrase: 'It is time to cut the tall trees'.[8] The Tutsis were very tall compared to the Hutu, and by the end of the machete massacre (which lasted for *100 days*), 800,000 Tutsis and Hutu sympathizers were dead.[9]

Imagine now a worldwide version of the Rwanda Genocide, where a signal goes out to those who are enslaved to the will of the Serpent through his servant, AntiChrist. Blood will run in every land, and believers—those who refuse to take the Mark of the Beast—will be on the run for their very lives.

The third horseman is next, are you ready? Come and see…

One Ring to rule them all,
One Ring to find them,
One Ring to bring them all
and in the darkness bind them.
—J. R. R. Tolkien

Nearly every fan of fantasy lore has the above quote memorized. It has been immortalized in Peter Jackson's trilogy of films and published again and again since its original 1954 printing. In *The Lord of the Rings*, Sauron is a type of the AntiChrist, and he wants to rule all of Middle Earth. But as dark and ominous as Tolkien's story can be even for adults, the truth is much more terrifying.

In the book of Revelation, the Risen Lamb, Christ the Lord is opening a series of seals, slowly opening up a legal deed to the Earth and its inhabitants. We turn now to Seal number Three:

> And when he had opened the third seal, I heard the third beast say, Come and see. And I beheld, and lo a black horse; and he that sat on him had a pair of balances in his hand. And I heard a voice in the midst of the four beasts say, A measure of wheat for a penny, and three measures of barley for a penny; and see thou hurt not the oil and the wine.
>
> —Revelation 6:5–6

This steed is black, *melas* in Greek, which means 'black as ink'. Later, in Revelation 6:12, this same word is used again to describe the state of the sun:

> And I beheld when he had opened the sixth seal, and, lo, there was a great earthquake; and the sun became black as sackcloth of hair, and the moon became as blood.
>
> —Revelation 6:12

The color black implies an absence of light, and it carries with it a sense of great doom. The black rider carries a pair of balances. The Greek is *zygos,* and it is generally referred to as meaning 'a scale' for weighing out grain or metals, but *zygos* embodies much more than just a reference to a scale. This word denotes labor, enslavement, and troublesome law for it refers to the 'yoke' put upon cattle. In contrast, Christ says He will exchange His light yoke (zygos) for our heavy ones. (See Matthew 11:29–30). The AntiChrist's yoke will be heavier than anyone can imagine, though his slaves may not realize it at first, for he will promise a libertine society and joy to all who take the Mark. He may even offer an upgrade to Humanity 2.0, but that is another book.

This balance, this *zygos* or yoke, is the One World Economy coupled to the New World Order's rule book, where none may buy or sell without the Mark. Imagine needing milk for your child, food, water, or those little comforts on which we all depend (toilet paper comes to mind), and then imagine being unable to obtain those without selling your soul in the bargain. What would you do?

But as dire as Riders One through Three are, the last, the Fourth Rider is even more terrifying, for we are talking about death. It is the ultimate end to all who call themselves humans, but it was not always so. In the beginning, Adam and Eve walked freely in the Garden of Eden, but with two simple rules: Do not eat of the Tree of Knowledge of Good and Evil, and Do not eat of the Tree of Life. God gave only two rules, and our distant ancestors just couldn't obey.

I could write volume upon volume about sin and why it exists. Most of us could. The whys, the wherefores, the witcheries—the deceit and conceit of that moment when the 'Serpent beguiled Eve and she did eat' forever altered human history. We are all born in sin because of it.

We are born without direct access to God.

Oh, didn't you know that? Death in Eden was instantaneous regarding Adam and Eve's relationship to their Creator—to God. Something inside them died—His spirit had departed. Remember, God literally breathed His Spirit into Adam and made him living creatures, but when Adam sinned, Adam changed. The knowledge within the 'fruit' itself did not so much alter the man as it was man's disobedience which did so. Our loving Creator God, however, intervened and provided a 'covering' for Adam's sin. The Father slaughtered an innocent creature—probably a lamb as a picture of our promised Redeemer, the Lamb of God Jesus Christ—and it was the *blood covering* of this sacrifice that formed the bridge that once again united Adam to God.

Death has been our enemy ever since, for not only did Mankind's spirit suffer change when Adam sinned, but he and all of us who follow suffer the same condemnation which includes physical death. Romans 6:23 tells us that "the wages of sin is death, but the gift of God is eternal life through Jesus Christ our Lord."

In Revelation 6, we meet Death, the Grim Reaper himself. Let's examine what John saw:

> And when he had opened the fourth seal, I heard the voice of the fourth beast say, Come and see. And I looked, and behold a pale horse: and his name that sat on him was Death, and Hell followed with him. And power was given unto them over the fourth part of the earth, to kill with sword, and with hunger, and with death, and with the beasts of the earth.
>
> —REVELATION. 6:7–8

Christ has been opening wax seals within the rolled scroll that only He is worthy to open, and now He comes to the Fourth Seal, and another horse gallops unto the Earth, bearing upon his back a dark, ominous rider. The most fearsome of all. And this rider is not alone.

With the first three seals, we've had a white horse, a red one, and a black steed the color of ink. Now, our Lord shows John a horse the color of sickly green—*chloros*. It is the color of new grass, just born, or the color of someone who is 'green to the gills', an old-fashioned phrase describing a person who is sick to his stomach, queasy.

But did you notice a major difference with this rider? The first three riders are merely referred to as 'he that sat'. No name is given to these three. The Fourth Rider, however, is *named*. He is called Death in the English translation, but the original Greek uses the word *Thanatos*. This is more than just a reference to death or 'loss of life'. This is the rider's NAME.

And Hell—*Hades*—is his companion.

John's contemporaries would have known their Greek mythology which tells of a supernatural entity named *Thanatos*, born of a union between Zeus and Nyx (night). I studied Classical Mythology in college, and I can tell you that there are hundreds of 'gods', so it is definitely not within the constraints of this short book to analyze the entire Greek and Roman pantheons, suffice it to say that mythology lists Chaos as the progenitor of Nyx, who joined with Zeus to produce Thanatos, the Angel of DEATH.

Zeus is one of many sons born to Cronus and Gaia. Cronus's name means 'to cut', and he and his sister Rhea were born in the first generation of Titans from the union of Uranus (the SKY) and Gaia (EARTH). In a pattern that is repeated by his son Cronus, Uranus hides two of his children from Gaia: The giants Hecatonchires ('hundred-handed') and Cyclops. Understandably, this makes Gaia very angry, so she asks her remaining children to help her remove their father Uranus from power. Only Cronus agrees, so taking a stone sickle provided by his mother, Cronus castrates Uranus and becomes ruler of the universe with Rhea as his Queen. However, bloody deeds beget bloody deeds, so believing his children would unseat him in similar fashion, Cronus decides he will *eat* his children before they can kill him.

Now, by the time Zeus was born, Cronus had already consumed all of baby Zeus's older siblings, and would have gulped down Zeus, except that Zeus had been hidden in a cave by his mother Rhea. There, Zeus was raised by his grandmother, Gaia (whom we might call 'Mother Earth'). Gaia probably told her son about his father's cannibalistic deeds and prepared young Zeus to exact revenge. So, once grown to adulthood and old enough to fight, Zeus tricks his father into drinking an emetic, which causes Cronus to vomit up his consumed children.

The myth of Cronus' castration of Uranus parallels the Hurrian Song of Kumarbi, where Anu (the heavens) is castrated by Kumarbi. In the Song of Ullikummi, Teshub uses the "sickle with which heaven and earth had once been separated" to defeat the monster Ullikummi, establishing that the "castration" of the heavens by means of a sickle was part of a creation myth, a cut that created an opening or gap between heaven (imagined as a dome of stone) and earth, enabling the beginning of time (Chronos) and human history.

Indulge me for a moment while I speculate. If, the Indo-European Kumarbi (Cronus to the Greeks) represents a primordial entity that created TIME, then the story about Cronus consuming his children may also have a time component. What if this story is really about a selfish god (read this as fallen angel) who imprisoned his children within himself—within TIME—that is within a time lock? If so, then Zeus forced his father not to disgorge his siblings but to release them from their time or even dimensional prison.

Ok, I'm speculating, but this mythic tale forms the foundation of a massive, world war known as the *Titanomachy*, or the War of the Titans. This war is said to have lasted for ten years and took place long before man was created. On the one side were the Titans, or the children of Uranus and Gaia—the original Giant Rulers. On the other side were the Olympians—essentially Zeus and his siblings. In case you're wondering, the Olympians won.

The list of the Twelve Olympians, also called the Dodekatheon, often includes HADES, although some claim he was too busy running the Underworld to care. Plato held that our twelve month time cycle corresponds to the Twelve Olympians, and that the last month is dedicated to HADES or PLUTO as the Romans called him. The Twelve are most often listed as Zeus, Hera, Poseidon, Demeter, Athena, Apollo, Artemis, Ares, Aphrodite, Hephaestus, Hermes, and either Hestia or Dionysus. Some believe that Dionysus is another name for HADES, although I cannot find strong support for this.

It is Zeus the Olympian who fathers Thanatos through Nyx. Thanatos's job is to escort all to the realm of HADES, land of the Dead, which makes Thanatos a lieutenant to HADES. Interestingly, Thanatos has a twin brother named *Hypnos*, also called Sleep, and it's said the brothers often travel together. The Apostle Paul is perhaps referring to Hypnos when he says that believers in Christ are not dead, but asleep:

> For this cause many are weak and sickly among you, and many sleep.
>
> —1 Corinthians 11:30

> Behold, I shew you a mystery; We shall not all sleep, but we shall all be changed,
>
> —1 Corinthians 15:51

> For if we believe that Jesus died and rose again, even so them also which sleep in Jesus will God bring with him.
>
> —1 Thessalonians 4:14

In Revelation 6, we see Thanatos riding upon a horse that is a pale green in color. Thanatos is the fourth rider to appear. He is given a *rhomphaia*, which is a curved, single-edged sword mounted upon a pole. This instrument of destruction is an important clue. A *rhomphaia* is similar to a scythe, the implement carried by The Angel of Death, and he can use this terrifying weapon without dismounting, allowing him to run swiftly through entire populations as the Lord permits, for remember that God has forbidden any creature to take the life of man—they may do so only with His permission.

In Job, we are told that Satan must seek authority from God to 'test' the faithful human, and God's answer is this:

> And the LORD said unto Satan, Behold, all that he hath is in thy power; only upon himself put not forth thine hand. So

> Satan went forth from the presence of the LORD.
>
> —Job 1:12

However, when Job responded by worshiping the Lord, even as he mourned the loss of his treasures and his children, Satan approached the throne once more:

> Satan answered the LORD, and said, Skin for skin, yea, all that a man hath will he give for his life. But put forth thine hand now, and touch his bone and his flesh, and he will curse thee to thy face. And the LORD said unto Satan, Behold, he is in thine hand; **but save his life**.
>
> —Job 2:4–6 (Emphasis added)

Job is an important book to consider when wrestling with the dark imagery in Revelation 6 and following. We can find comfort in knowing that all those who love Christ, who are called by His name and are covered by His redeeming blood, rest in His protecting hand, so nothing reaches us that is not permitted by Him for His reasons and for His glory. So it was with Job, and so it will be with those who are targets of Satan's ultimate test.

Remember what Satan says to God, "All that a man hath will he give for his life."

Thanatos and his pal Hades will put that to the test in the very near future. What are Death's weapons against us? How will he harvest mankind with his weapon, the rhomphaia?

We'll explore that next. Come and see…

So went Satan forth from the presence of the LORD,*
and smote Job with sore boils from the sole
of his foot unto his crown.

—Job 2:7

He saw virus particles shaped like snakes, in negative images. They were white cobras tangled among themselves, like the hair of Medusa. They were the face of nature herself, the obscene goddess revealed naked. This life form thing was breathtakingly beautiful. As he stared at it, he found himself being pulled out of the human world into a world where moral boundaries blur and finally dissolve completely. He was lost in wonder and admiration, even though he knew that he was the prey.

—Richard Preston, *The Hot Zone*

When the topic of Ebola comes up, there is one book that is always mentioned, *The Hot Zone* by Richard Preston. Part biography, part history, part biotech thriller, *The Hot Zone* tells the true story of the men and women who uncovered the first Ebola outbreak on U.S. soil. Here is where I learned much about the bravery and sometimes foolishness of those who stand on the frontlines of the battle with hemorrhagic diseases. And it was in those pages that I first read how scientists refer to the Ebola virus as a worm or a 'snake'.

Ebola and her sisters began to bite just about the time that Israel was forming into a modern nation, with the earliest known hemorrhagic fever disease outbreak in South Korea in 1953. Some point to an incursion in 1933, but that is uncertain. Either way, the virus is a deadly enemy, and it is a weapon granted to the rider Thanatos.

In the previous chapter, we discussed the nature of this dispatched rider called Death, and we examined his deadly *rhomphaia*, which is shaped not unlike the Ebola and Marburg viruses: A linear form that ends in a hook or curve. But notice that Thanatos is given permission to kill with sword (*rhompaia*), famine (*limos* in the Greek, meaning scarcity of harvest), death, and the BEASTS of the Earth.

When I first encountered this word 'beasts' as a novice Bible scholar back in my junior high and early high school years, I remember picturing great lions and bears ripping men apart. However, as I've grown older and studied biology—since I've spent long hours peering into the microscopic world of tiny beasts, the picture in my mind has shifted from one of massive lions to that of tiny invaders that man cannot see with the naked eye.

In fact, *therion*, the term used here, is a Greek diminutive of '*ther*', which is a figurative term for 'preparing destruction of men'. '*Ther*' is translated in Romans 11:9 as 'trap'. Curiously, because I am fascinated by numbers in the Bible, the other book that begins with 'R' is Revelation, and Revelation 9:11 (the opposite of 11:9) names the King of the Bottomless Pit as Apollyon or Abaddon. Is it conceivable that the little trap of *therion* (the tiny beasts) will help to set up the rule of Apollyon?

Again, I may be reaching, but in my biotech thriller *The Armageddon Strain*, that is exactly what happens. The plot is this: The panic caused by a worldwide pandemic caused by a laboratory created H5N1/Ebola hybrid leads to the acceptance of a universal nanoscale chip called the BioStrain Chip, but this is all a ruse to set up the Beast System of AntiChrist; it is a *trap,* and likewise, the real diseases unleashed by Thanatos in the near future (with Ebola as a harbinger of those *therion*) may be a cunning deception permitted by the true Ruler of the Universe, Almighty God.

Disease has been a part of mankind's life since the fall of Adam when Death was first given permission to take mankind. At that moment, God also cursed our World. In Genesis 3:17, God tells Adam that the very ground was now cursed because of his sin. In verse 18, God declares that the Earth would henceforth bring forth thistles and thorns, and Adam would have to scratch out his living—producing food for his family—from a cursed Earth by the sweat of his brow and the use of his hands. No longer could he and Eve enjoy the fruits in God's pleasant garden, but they must eat bread made by their own toil until the day they died.

Dust they were, and unto dust they would return.

Not long after this stunning pronouncement upon the sinful pair, the Earth experienced a population boom, filling the land mass with sinful people, and by the sixth chapter of Genesis fallen angels had assumed control—possibly some of those sent by God to keep an eye on His human children, those known as Watchers. These supernatural beings looked upon the human women and found them beautiful.

Genesis 6 tells us that the offspring of these unions, whether consensual or through rape, were giants, the Nephilim. Angels, and perhaps also their Nephilim children interbred with humanity for generations through conventional sexual means. However, since the ultimate goal of this hellish eugenics program was to prevent Messiah's birth, it is conceivable that the Watchers also employed scientific means to alter mankind's genome.

DNA manipulation would have been simple to the Watchers. They could have used recombinant and even epigenetic methods that humanity has only now discovered to create genetic chaos in pre-flood humans. But they must have known that God would intervene. The enemy is always trying to outwit God, and I believe that the Watchers may have used their knowledge to create a failsafe method to continue acting upon man's DNA for millennia to come by building microscopic machines.

Today, we call these biological machines viruses. Ask any virologist, and he or she will tell you that virions (virus particles) are not really alive. They have no metabolic functions. They merely hijack their host's metabolic processes to make copies of themselves.

Even though they are not considered living material, viruses are classified according to Order, Family, Genus, and Species. There are seven orders, ninety-six families, twenty-two subfamilies, four-hundred-twenty genera, and two-thousand-forty-six species currently listed in the catalog of virion particles. Of the ninety-six families, the one that most terrifies today's virologists is the family *Filoviridae.*

This family has only three offspring or 'genera': These are the Cuevavirus, Marburg Virus, and Ebolavirus. The first, Cuevavirus (Cueva is the Spanish word for 'cave') is more like a distant cousin than a sister to the other two, for it differs from Marburg and Ebola by more than fifty percent of its genetic material and, to my knowledge, has never engendered disease in humans.

Ebola is another, younger sister of the most lethal of all hemorrhagic fever viruses, the elder sister being Marburg. In 1967, this older microscopic 'serpent' machine emerged from its secret hiding place to strike the city of Marburg, Germany. There, the Behringwerke Pharmaceutical plant routinely imported monkeys whose tissues were used for testing the efficacy of new drugs. However, one ill-fated group of African grivets arrived at the facility that year carrying more than just fleas or common simian illnesses. These monkeys had a secret payload, a virion previously unknown to medical science, a serpentine virus with a simple, negative-strand RNA genome with only seven, deadly genes.

During that warm August, workers in the Marburg plant went about their jobs of slaughtering the monkeys to recover kidneys, which at that time were commonly used in preparing growth media for culturing vaccines. Before Thanatos was finished with his ride, 31 people were dead in Marburg and neighboring districts. The lethality of Marburg, the blood serpent, was later determined to be as high as *ninety percent.*

Marburg and Ebola have ravaged through many communities unlucky enough to discover them—primarily in Africa. These hemorrhagic fever viruses, these ancient machines, these *therion* tiny traps, have a unique and distinctive appearance, a linear shape with a hook or even a 'cheerio' or circle at the end. Some call this the 'eyebolt'. They also resemble ropes, worms, or serpents coiling upon themselves, which is appropriate considering that Watchers (those fallen enemies of God) may have created these genetic copy machines.

Marburg and Ebola are amazingly simple. They contain only SEVEN genes. Since 1967, Marburg has emerged eleven times with endemic outbreaks and/or incidents (an incident is a lab accident). These have occurred primarily in central Africa, but also in Europe and Russia—and the United States. The death rate varies from about twenty-five percent to as high as ninety percent. For comparison, the death rate for the Spanish Flu was between two and five percent.

Ebola officially emerged in Zaire in 1976, and since then it has made eastern Africa its home, presumably using bats or another indigenous species as home base.

Why did it suddenly appear in 1976?

Earlier, we looked at Christ's message to His disciples as record by Matthew in chapter twenty-four of his gospel. Jesus warned them that 'the end of age' would come with increasingly frequent and intense signs such as pestilence, plagues, earthquakes, and war. The twentieth century saw these signs increase in frequency and intensity, opening with the first true, World War, followed by a series of fatal and debilitating plagues that afflicted all the human populations of the world, and in particular Africa: Spanish flu, polio, smallpox, yellow fever, tuberculosis, cholera. Some say AIDS began as early as 1959 when the oldest sample of HIV-1 was isolated in Kinshasha,

Zaire, the same area where Ebola emerged in 1976. This HIV sample is said to have come from a Bantu male. Another from an adult Bantu female was isolated in 1960 and shares approximately eighty-eight percent genetic similarity to the 1959 strain.

Though AIDS is not the primary topic for this discussion, it is important to note that the official timelines for HIV and Filoviridae commence at about the same time, the era when African colonies were gaining their independence.

It was in East-Central Africa in 1976 that Ebola emerged in two different areas simultaneously: Zaire (now called the Democratic Republic of Congo) and Sudan. The years leading up to 1976 had been filled with turmoil and heartache for the peoples living in the Uganda/Sudan area. Beginning in 1962, when Uganda gained independence from Great Britain, a series of despots ruled the region. First Milton Obote, then Idi Admin ruled as if Uganda were their personal playgrounds. Persecution, famine, rioting, and rebellion ravaged the economy—such as it was—and stripped it to nothing, leaving the people of the area destitute.

Women, in particular suffered. With many of their husbands working in the cities, these half-starved wives and mothers beat the earth to scratch out food. The wars left many women completely alone, and these survived the only way open to them in a nation left economically bankrupt—by selling their bodies, working in the cities and along the Kinshasha Highway as a prostitute.

The beautiful, mysterious continent that had been surveyed and mapped by Dr. David Livingstone had succumbed to European invasion and been conquered. Africa was suffering the Braxton

Hicks birth pangs, harbingers of the white horse and rider, who will soon come conquering and subduing; of the second Rider who will bring war; and birth pangs of rider number three who will very soon bring economic collapse to the world.

After this Africa felt the sharp and warning pangs of the fourth rider; future echoes of Thanatos, who will bring plagues to those who survive the first three. As if to show the world what he is capable of, Thanatos swept his *rhomphaia* across the backs and necks of desperately poor, half-starved Ugandans and Sudanese and by 1976, a strange new virus emerged from the dense jungle.

More on this next. Come and see…

In biology, nothing is clear, everything is too complicated, everything is a mess, and just when you think you understand something, you peel off a layer and find deeper complications beneath. Nature is anything but simple.

—Richard Preston, *The Hot Zone*

Only the dead have seen the end of war.

—Plato

The African state of the DRC (formerly Zaire) shares a common tri-state border with Uganda and South Sudan, so it's not unusual for a man or woman, particularly one earning a living along the highways, to travel back and forth. By 1976, war and famine redistributed not just humans but also animals. Famine engendered the need for more farmland, so large swathes of jungle came down so the land might be tilled. Animals, particularly rodents, follow the path of grain, and rural Africans kept their precious stores of grain safely hidden inside their huts. This meant rats and mice, hungry for grain, multiplied by the hundreds beneath granaries, sleeping during the daytime and traveling amongst their human hosts at night.

And so it was that one day in September 1976, Mabalo Lokela began to feel feverish. As with many of his friends and family, Mabalo, known to his friends as Antoine, had suffered with malarial fevers and knew just what to do because he taught school in the Mission. Yambuku Mission compound included the Bumba Zone's only hospital. Whenever the fevers and chills would return, the kind Belgian sisters would give Antoine an injection of chloroquine, and he would be right as rain. To Antoine, and to most of those living near the hospital, needles meant cures. But this time would be different.

Each day, as many as four-hundred patients would line up for medicine and shots in the tiny missionary hospital. Money and supplies were always scarce for the sisters. Gangs and corrupt officials often hi-jacked precious medical supplies and sold them for profit. This meant the hard-working, hard-praying nurses of Yambuku had to re-use their precious needles, sometimes forgetting to clean them properly between jabs.

Every morning, the sick, many of whom had walked for hours to get there, lined up for medicine, and pregnant women came for the miracle Vitamin B shots that gave them energy so they could keep working. Along with the nuns, barely trained novice nurses did their best to practice cleanliness, but with only five hypodermic needles[10] and hundreds of arms, blood from one person often infected another.

Once he'd received his miracle chloroquine shot, Antoine returned to his home village of Yalikonde for a rest. Two days later, on August 26, 1976, a strange man came to the tiny hospital complaining of terrible diarrhea. The stranger said he was from Yandongi, a nearby village in the Bumba Zone. The sisters put the man into one of their 120 beds and decided to treat him for dysentery though he also suffered from extreme nosebleeds. After two days with the sisters, the stranger suddenly left—he was never seen again though later, many WHO and CDC researchers would trek from village to village trying in vain to find him.

Poor Antoine was getting no better, so he checked in again with the nurses but was told he needed more rest at home. On September 5, 1976, he returned complaining of violent diarrhea and looking so dehydrated that the nurses said he had 'ghost eyes': deeply recessed, glazed, and surrounded by pale skin. And he was bleeding. His gums bled. His nose bled. His vomitus contained blood, as did his diarrhea.

Soon other patients, many of them women who had received vitamin injections, began to straggle into the Mission hospital, bleeding from their noses and staring at the nurses with ghost eyes.

Three days later, on September 8th, Antoine died. And soon others followed in death, and their family members contracted the bleeding illness, their eyes turned ghostly, and they too died. The tiny hospital filled with panicked, bleeding villagers. Twenty-one of Antoine's friends became ill; eighteen died. The sick often grew delirious, tearing their clothes from their bodies and running as if to escape some unseen threat. Perhaps they could see the spirit of

Thanatos wielding his terrifying weapons *rhomphaia* and *therion*, the small but deadly Blood Serpent.

Before the disease finally burned out, 318 people had contracted the disease that Peter Piot would later name Ebola after the nearby river. Out of that 318—280 died. An eighty-eight percent fatality rate.

At nearly the same time, Ebola also broke out in Sudan, killing 151 out of 284—fifty-three percent fatal. Since 1976, there have been many more Ebola epidemics in central Africa, but each outbreak spared the western coast of Africa until now. I'll get to the current Ebola epidemic in a moment, but first let me tell you about a small outbreak that barely made the news, even though it was discovered and documented in a bestselling non-fiction work by Richard Preston. And even though Preston's account would even inspire a film, I would wager that most Americans are unaware of how close our nation came to becoming a nation of ghost eyes.

The Hot Zone tells the story of an uncomfortably close foreshadowing or 'birth pang' of Thanatos and his attempted ride through Reston, Virginia. He arrived in 1989, hidden within a shipment of crab-eating monkeys. The monkeys had originated in the jungles of Manila, captured by Ferlite Scientific Research Farms who sold their imprisoned simians to labs in the US. However, before these monkeys were permitted to continue to their final destinations, US customs required that the monkeys be quarantined for thirty days in a qualified holding facility.

And so it was that on October 2, 1989, one-hundred crab-eating macaques (a type of monkey with long, very sharp fangs) flew from Manila to Amsterdam via Tokyo and Taipei. They left Taipei, arrived in New York City, and then made their way via truck and highway to Reston, Virginia, joining five-hundred other macaques in a temporary home at the Hazelton Facility. Considering the stress the monkeys endured in transport and making allowances for some indigenous illnesses, it was not unusual for some to arrive dead or to die soon after, but when one-hundred monkeys died, Hazelton veterinarian Dan Dalgard began to worry.

Dissection of the bodies revealed a large volume of blood throughout the body, especially in the lungs, but the spleen looked like a large, rock-like mass. Normally, a spleen is like an underinflated balloon filled with blood. Squishy, not hard.

Puzzled and a bit worried, Dalgard decided to call in the big guns. USAMRIID, the US Medical Research Institute of Infectious Diseases, was the only lab capable of diagnosing a Biosafety Level 4 virus. Filoviruses such as Ebola and Marburg are BSL-4 contagions. This means only specially trained scientists wearing positive pressure suits, following very strict guidelines are permitted to work with them. Dalgard's presumption early on was that the monkeys had died of Simian Hemorrhagic Fever, so while USAMRIID ran their tests, the small team at Hazelton began euthanizing sick monkeys, followed by necropsies on all the dead, which meant the vets and assistants were up to their elbows in lots of infected blood.

The initial batch of sick monkeys were housed in Hazelton's Room F, but monkeys in other locations in the building soon began to exhibit symptoms. Ebola and other hemorrhagic fevers are generally spread through direct contact—eg. via a contaminated needle or through contact with bodily fluids. In Africa, families traditionally wash their loved ones' bodies and clean out all body cavities in preparation for burial, which is why so many family members became sick. But how did monkeys in one room infect monkeys down the hall in a separate, locked room—with no direct contact?

At first, it was thought that facility workers had not observed correct sterilization procedures and carried the virus with them from room to room, but it soon became clear that the virus had traveled through THE AIR DUCTS. In fact, several HUMAN workers also became ill, but only mildly so. Blood tests would later confirm the presence of Ebola antibodies in the bloodstreams of these workers, and it was determined that these men had very likely become infected via contaminated surfaces or even via Ebola-laden droplets in the air.[11] Because Ebola Reston was contracted through the air ducts, the infected monkeys outside of Room F had taken the virus into their lungs, where it had multiplied and exited through coughing.

It was in Reston that a shocked team of CDC and USAMRID doctors realized that Ebola CAN TRAVEL through the AIR.[12] However, let me be clear: So far, this is a weak infection route for humans. (See the endnotes[13] for more on airborne and aerosolized transmission).

Years later, CDC researchers discovered that many of the jungle workers in Manila, those who routinely catch and handle crab eating monkeys, have Ebola antibodies in their bloodstream. Ebola may not be currently efficient at aerosolized infection, but one tiny shift in its seven genes, and human to human transmission rates could increase exponentially.

Ebola slept for years, but in 2014 it woke in a new region, and has dominated news cycles for most of the summer and now into the fall.

We'll discuss the West Africa outbreak next.

Come and see…

"I know," said Butuzov. "We were thinking of something like Ebola."

"That would work, but you'd run the risk of killing everyone around him."

"That wouldn't matter."

—Ken Alibek,

quoting a conversation with a fellow bioweapons researcher in his book *Biohazard*

Ebola virus is a microscopic trap. It waits for its victims. It is patient. The 2014 outbreak may only be a birth pang (and let us pray that it is), but if projections prove true, by 2015 Ebola infections will skyrocket.

Earlier this year, Ebola suddenly emerged in West Africa for the first time. To the shock of all Ebola researchers and to the utter dismay of those affected, Ebola was confirmed as the cause of a series of hemorrhagic disease deaths in March. As of this date (October 9, 2014), Ebola is running rampant in Guinea, Liberia, and Sierra

Leone, and has now emerged in Nigeria, Senegal; and in east Africa, the birthplace of Ebola, the DRC (Democratic Republic of Congo) is also reporting cases. However, according to the CDC, the DRC outbreak is not directly related to the West Africa outbreak.

But we have also seen cases in the United States and Spain. On October 8, 2014, Thomas Eric Duncan, a Liberian immigrant who became the first case to be diagnosed with the borders of the United States, died.

The CDC website lists the current total 'case' count at 7,492, and the total death count at more than 3,400 human souls, but each new day brings news of more dead. The World Health Organization, the CDC, and Doctors without Borders are responding, but officials from these organizations have told reporters that the outbreak is out of control[14]. Projections by the World Health Organization released in September indicate that at the present rate of infection, there could be 1.4 Million cases by end of January 2015. Of that number, half would be dead.

The West African version of Ebola Virus has been determined through genetic analysis to be ninety-seven percent similar to Ebola Zaire from 1976, but one has to wonder if the tiny changes have made it deadlier. A global recession and threats of war across major theaters had—until now—dried up donations to WHO and Doctors without Borders, making it more and more difficult for the teams to adequately respond.

A recent interview with members of the CDC team in West Africa illustrates the desperation these men and women feel when faced with such a daunting task and so little funding:

> "The situation in West Africa should be a wake-up call to recognize that this weakening of this institution on which we all depend is not in anybody's interest," Scott Dowell, director of disease detection and emergency response at the U.S. Centers for Disease Control and Prevention, said during a briefing in Washington. "In my view, there's **no way that WHO can respond in a way that we need it to.**"[15] (Emphasis added)

In early July, it was announced that the physician who has been leading the charge against Ebola in West Africa has contracted the deadly disease. Thankfully, due to the excellent healthcare provided for him in controlled surroundings as well as the administration of the experimental drug ZMapp[16], Dr. Kent Brantley has been discharged in good health and hopes to return to Liberia to fight the good fight.

When I presented my Thanatos talk in July at the Pike's Peak Conference, I warned the audience that Ebola was just a plane ride away from nations outside Africa. Just two weeks ago, mainstream media began to shout in large, red letters across the Internet that a patient had tested positive for Ebola in Dallas, Texas.

The Ebola virus is tiny—as small as 900 nm. That's 0.000000097 meters. A typical strand of hair is from 50-100 microns in diameter. Ebola then is about 1/50 to 1/100 as thick as a human hair. That is really small.

Infection can occur with only 5-10 virions (virus particles). Teeny tiny. A drop of blood from an infected individual can contain 100 MILLION virions. Scared yet?

I'm amazed at the brave men and women who volunteer to work in Ebola-stricken regions such as Liberia. Oftentimes, these wonderful people represent Christian organizations like Samaritan's Purse (eg. Dr. Kent Brantly mentioned earlier and nurse Nancy Writebol, who also recovered from Ebola infection stateside), or the Catholic nuns who treated Ebola Zaire patient Antoine and many others. Most of the time, however, the volunteers or workers who treat victims have little training and live near or in the treatment center. They all know and understand the high risk one takes when handling fluids that writhe with the Blood Serpents, but they do it. They simply get on and do it.

Ebola infection begins when a few or many virus particles enter your body. The next step in what's called pathogenesis (the process of infection) depends on the route of infection. Most victims are infected when caring for a loved one or handling the dead as bodily fluids such as blood, saliva, vomitus, feces, or even tears enter through a tiny cut or break in your skin, your mouth, your nasal passages, or your eyes. Ebola virions also travel in semen or vaginal secretions, so intimate moments with an Ebola victim (perhaps one not yet showing symptoms) may result in a deadly infection.

Sometimes, the route of infection is parenteral. This means, the infective agent is injected into the body. You might rightly ask, who in their right mind would do that? My sister, who is a nurse, certainly did when we were discussing this route of infection earlier today (she was nice enough to ask how the new book was going—thanks, Deb!). And she's right to ask—as are you. Who in their right mind *would* do such a thing?

Earlier I related the story of Mabalo 'Antoine' Lokela who became Patient Zero in the Zaire outbreak in 1976. He taught at a mission school, where he grew to trust in the medicine of the Belgian sisters. It was in this hospital that many of the patients and local citizens became ill within a short period of time. Later, after examining what had actually happened to escalate the infection rate, WHO researchers learned that the nuns in the poorly equipped and cash-strapped mission had very few hypodermic needles, so they routinely washed and 'sterilized' them between uses.

In her sobering book, *The Coming Plague: Newly Emerging Diseases in a World out of Balance*, Laurie Garrett reviews what happened at the mission hospitals that resulted in so many deaths:

> Both in Maridi and in Yambuku the poorly supplied clinics reused syringes hundreds of times a day, injecting drugs from one person to another without sterilizing needles. McCormick calculated that during the months of September and October, 1976, an individual's odds of getting Ebola virus from a single injection at the Yambuku and Maridi hospitals *exceeded 90 percent.* [...] Sureau calculated that 43 percent of the Yambuku area Ebola victims who got the disease from another person survived the ailment, but only 7.5 percent of those who were injected with contaminated syringes survived.[17] (Emphasis in original)

At the Yambuku mission hospital, local women who were pregnant had become accustomed to weekly injections of vitamin B complex to give them enough energy to work and provide for their children and husband. These injections carried a deadly payload, though no one knew it. And a pregnant woman with Ebola nearly always miscarried and bled, and bled, and bled.

If you Google the terms Ebola and bioweapon, you'll find (at least on my browser) 89,200 hits. Many of these are mainstream media articles that seek to assure the public that there has never been a laboratory-induced, an intentional mutation, or genetic alteration to Ebola. That is simply naïve. The Biopraparat Program and Vector Programs in Russia that were supposedly dismantled in 1972 when 170 nations signed the Bioweapons Treaty continued to research pathogens under the banner of 'defense'. The United States signed the 1972 resolution, and our country has also continued to research pathogens and their mutations in the name of 'biodefense'.

Personally, I would love to believe that both countries have been genuine in their statements and that neither has any plans to use their research to harm any humans, but as I said—that would be naïve. In his non-fiction expose, *Biohazard: The Chilling Story of the Largest Covert Biological Weapons Program in the World—Told from Inside by the Man Who Ran It,* Ken Alibek reveals biotechnical intrigues that will curl your hair. Seriously, if you want to sleep well, do not read this book just before bedtime.

Among the diseases that he lists as being in the top favorite pathogenic targets for weaponization are Ebola and her sister Marburg. I know that there are loads of theories across the Internet that the current Ebola outbreak is either a hoax, or that it's not a natural outbreak—that this is an engineered virus that has been genetically altered to make it truly airborne and aerosolized, but I will *not* tell you that. What I will say is that Alibek claims Russia engineered a *vaccinia* virus with an Ebola gene inserted, which theoretically would be transmissible via aerosol. (An aerosol transmission means that virus particles are *suspended* in the air and remain there for sufficient time to infect persons who are not directly in front of, say, someone who sneezes infected 'droplets' into the airspace).

Here is what Alibek claims:

> Our scientists had found it more difficult to cultivate Ebola than Marburg—they were not able to reach the necessary concentration, but by the end of 1990—the long-term problem of cultivation had been solved, and we were close to developing a new Ebola weapon.[18]

My purpose in stating this is to make you, dear reader, aware that there is strong potential that Ebola has already been genetically weaponized. Of course, **this does not imply that the West Africa outbreak is anything but natural.** But with so many terrorists in the world wanting to find a way to demoralize and bankrupt western nations, and considering that laboratory security is not always as 'secure' as one might hope, then isn't it logical to propose that one day, someone WILL release a bioengineered pathogen such as Ebola?

Try to sleep tonight, if you can. I know that my own sleep has been rather difficult while writing this.

So, how would an outbreak begin in the US? That's up next. Come and see....

With the framework of current U.S. government guidelines, such as the National Strategy for Countering Biological Threats and IHR (2005), the world is closer than ever to truly working together on surveillance and control of infectious diseases without consideration of borders.

—Kevin L. Russell,

et.al "The Global Emerging Infection Surveillance and Response System (GEIS), a U.S. government tool for improved global biosurveillance: a review of 2009"[19]

So, let's say that someone in your hometown is diagnosed with Ebola. Where we live that is a very real possibility, since we live in a college town, and international students come and go freely. It is certainly in the realm of possibility for one of these students to bring a hitchhiker along in their bloodstream. The incubation period for Ebola is from two to twenty-one days, so the likelihood of someone traveling and being permitted to travel (being asymptomatic) is high. Extremely high.

Let's examine the possible timeline of a fictional student teacher named 'Mary' who returns from a job interview in Paris, France:

—

Mary arrives home in her small, college town, (on **Day 1** after exposure), she returns to student teaching at an elementary school a few blocks from your home. Mary feels fine then, a little jet-lagged, but all right nonetheless. Coffee and the odd donut fuel her with enough energy to make it through the three days left in her week, and she comes home on Friday—**Day 4**—feeling weary but determined to go to the fall festival at her college.

Mary returns to teach on Monday (**Day** 7), but she is achy and sneezing. She assumes she has caught a cold in France. During this week and a half of teaching, Mary touches and hugs and sneezes around fellow teachers, fellow college students, and her class of twenty-three second graders.

By Friday, **Day 12**, Mary decides to stay home, and her mother encourages her to see their doctor or at least report to the college clinic. Mary's mother has been reading about a patient in Shrewsbury, Connecticut who is being tested for Ebola and another in London, but she has no reason to fear for her daughter, who'd not been anywhere near West Africa.

Mary's mother knows that her daughter traveled abroad but doesn't realize that one of the passengers on Mary's plane was shedding virus, and Mary caught it when she used the restroom area after the Ebola-infected unknown person. (Touching surfaces that have been infected can spread Ebola—the virus can remain viable for several days.[20])

Day 14 arrives, and the news is buzzing about the London man who has tested positive, and a new individual—an elderly grandmother from Brussels who worked in the airport there—who has tested positive. Mary is beginning to run a fever, so she calls her doctor, but he's booked up. That night, Mary begins to experience a sharp throbbing in her head, and she's a bit dizzy. Her stomach hurts a bit, and she assumes she's picked up a bug from one of her elementary school students.

The next morning, on **Day 15** since she left Paris, Mary's mother cannot rouse her daughter from sleep. Mary now has diarrhea, and she has begun to vomit. She can barely move. Her eyes are red, and her throat feels like it's on fire.

Mary's mother is convinced that her child has something more than 'a bug', so the pair head to the emergency room at the county hospital, where the on-call resident diagnoses Mary with intestinal flu, sending her home with pain killers, a prescription for an antibiotic, and the usual advice to drink plenty of fluids and rest.

Day 16, and Mary exhibits hallmark, Ebola symptoms: diarrhea, severe stomach pains, violent vomiting, and her eyes are bloodshot. Her vomitus contains bright blood, which alarms her mother. Mom calls the hospital. She's told to bring her daughter in right away. The assumption by the physician on call is that Mary has a perforated ulcer. He admits her.

Day 17, Mary is lethargic, and her diarrhea is now constant and bloody. She continues to vomit, and she is feverish and suffering from bouts of chills and mild delusions. The doctor asks if she has visited any African nation, but Mary reports—truthfully—that she has not, however she was in France. Though he believes the chance that Mary has contracted Ebola is small, the doctor rushes a sample of her blood to the CDC labs in Atlanta. Recent news reports have caused a deluge of samples to land in the CDC offices, and it is four days before Mary's samples are processed. She has tested positive. She has Ebola.

Day 21: Mary is nearly comatose, and her capillaries have begun to break down. She is bleeding from the nose and eyes, and any injections of morphine or the new Ebola trial medicine from Glaxo-Smith-Kline cause blood to spout from the injection site like tiny, crimson fountains. Her skin is darkening from blood that has pooled just beneath the dermis.

Day 24: Mary is dead. She is officially the first person to die of Ebola in the United States.

This fictional scenario is presented to help the reader understand the wide and branching tree of contacts an Ebola patient creates. Mary contracted the disease from a man who should never have been allowed on the plane. You might argue that no person who is shedding virus would ever be permitted to fly, but that is not necessarily true. Currently, airport screeners look only for elevated temperatures, but not all Ebola patients register as feverish. Some may have low-grade fevers, or none at all. Screening also assumes that thermometers are accurate, and that those using them are consistent and efficient.

Airports in the Ebola stricken region of West Africa tell WHO and the UN that they monitor temperatures along with asking each passenger to answer a series of questions regarding their contacts within Ebola areas. Did they have contact with anyone exhibiting symptoms? Did they touch an Ebola patient? However, current measures are not foolproof, because not everyone will answer these questions truthfully.

Case in point, is Thomas Eric Duncan, a Liberian male who recently tested positive for Ebola in Dallas, Texas. Duncan was not completely truthful when answering questions at the airport in Liberia, claiming he had not touched anyone with Ebola nor come into contact with an Ebola patient, when in truth he had ridden in a taxi with and actually carried a pregnant woman with Ebola several days before traveling.

In our fictional case, someone must have been, let's say, disingenuous when answering his questionnaire. Sometimes, people with money can bypass certain safeguards. Humans often panic when stressed, and a fearful person is prone to make disastrous choices. In the case of Ebola or an agent similar, one disastrous choice may be all it takes to infect an entire nation.

Our fictional patient Mary's direct tree of contacts as she developed symptoms and began to shed virus would have included her mother, any family members whom she visited, her college friends and classmates, her students at the elementary school, people on public transportation, strangers in grocery stores, fellow believers at church, her healthcare contacts, and so on. But that tree continues to branch into indirect contacts, because if even one of those contactees becomes ill, then everyone he or she 'touches' in life is a potential victim.

Derek and I live in a small, mid-Illinois town, and I work from home, so I don't see many people during the day, but he travels throughout the state as an outside salesman. He shakes dozens of hands a day all across a large region. If Derek were ill, say with influenza, he could easily spread that illness to a hundred people in one day, who would then spread that to their contacts, who would share the virus with theirs—and it goes on, and on, and on. This is how pandemics are built.

In compact, urban areas the chance for contact rises substantially. Liberia's capital Monrovia is densely packed, and the death toll in that city alone as of this writing is over 1,300. WHO estimates, however, that the reported numbers are low by a factor of three, so it is entirely possible that the real death toll in Monrovia is nearly 5,000. If the transmission rate does not slow, then deaths will escalate exponentially very soon. Now, substitute Monrovia with Cleveland or Kansas City, or Chicago—or your home town.

This morning, I read that the Hajj is nearly over in Saudi Arabia. This annual pilgrimage made by Muslims all across the world brings believers to the holiest site in their entire belief-system, Mecca. BBC News, reports[21] that two million faithful performed the Hajj this year, arriving from their diverse countries, and who plan to return to their homes within the next few days. Wisely, Saudi Arabia would not permit anyone from an Ebola outbreak country to make the

Haj, but there is always the possibility that someone broke this rule. With two million people packed closely around the Kaaba (the elaborate cubical house that encloses the 'rock' that fell to Earth, probably a meteorite that holds special significance for Muslims), sharing a virus of any kind becomes the equivalent of sharing a ticking time bomb.

As an aside, one other very real problem that Saudi Arabia must solve is the current MERS (Middle Eastern Respiratory Syndrome) threat. So far, this is a relatively uncommon disease, but it is spread via aerosolized droplets in the same way SARS (Severe Acute Respiratory Syndrome) was spread in 2003. MERS is a very real threat to the Middle East, and potentially to the world, so we must keep an eye on that developing situation as well.

Adding to the disturbing aspect of airline-created infection 'trees' is a leaked report that claims 3,500 passengers from Ebola stricken nations have entered the US since January 1, 2014.[22]

Here is an excerpt from a Breitbart news article about the leaked report:

> The leaked report specifically reads, "According to CBP [Customs and Border Protection] data, since January 1, 2014 to June 30, 2014, a total of 3,566 passengers with a nexus to Guinea transited through or arrived at U.S. airports." The term "nexus" refers to passengers who flew from the Ebola stricken nation to a second nation, and then from the second nation into the United States. Guinea is attributed as the nation of origin for the current Ebola outbreak.

> The lack of any special processes or testing for individuals with a nexus to Ebola affected nations is illumined in the leaked internal report as well. It reads, "The Level 3 travel alert issued by the CDC on July 31st remains in effect as of August 13, 2014. The travel alert urges all US residents to avoid nonessential travel to Guinea. Although CBP is not doing any additional screening of passengers from the affected countries, CBP has enhanced their screening routine processes through guidance and training. Additionally, CDC is providing assistance with exit screening and communication efforts in West Africa to prevent sick travelers from boarding planes."[23]

The leaked report presents the very real possibility of exposure to a pathogen while on an airplane, or exposure at an airport, but more to the point it substantiates the example given of our fictional victim Mary and the branching tree of contacts issuing from just one exposure.

Scared now? There's more on pathogenicity. Come and see...

By March 6 Pinneo could longer eat or swallow because her throat—
the esophageal lining—was aflame with florid infection.
The worried medical staff noticed that Pineo's face and neck were
swelling. Her lung and chest were also filling with fluids, and X-rays
showed that some organism had invaded the linings of her lungs.

—LAURIE GARRETT, *THE COMING PLAGUE: NEWLY EMERGING DISEASES IN A WORLD OUT OF BALANCE*

"I ran some tests of my own. They have done something to us
that may affect everyone. You understand this? The blood tests
indicate the presence of genetically altered Ebola virus."

—FRANK BLACK TO PETER WATTS, MILLENNIUM EPISODE "THE FOURTH HORSEMAN"

When a Blood Serpent like Ebola strikes, the victim generally has no idea of the event. Only when infected during obvious routes such as needle stick or being covered in a known victim's fluids is the new victim aware that Ebola has wormed its way into his or her body. In most cases of infection, the advance of the Serpent's poisonous activity is slow and steady. However, if the route is via a needle stick, then the victim may show first symptoms within as little as a day.

It takes just 5–10 viral particles to bite you. Of course, when I speak of Ebola as a 'blood serpent' this is a metaphor. The virus may have a snake-like appearance, but it does not directly bite. Instead, like a subtle trickster, the virus fools the human immune system into believing it is 'friendly' while simultaneously releasing signals that induce what is called a fatal 'cytokine storm'.

Ebola virus particles (also called virions) can infect almost any human cell, but they have a special affinity for connective tissue—eg. collagen and blood. A viral particle contains an exterior 'skin', called the viral envelope, which has small protrusions along the outer surface known as glycoproteins (GPs). This viral envelope looks friendly to human cells, because it is derived from previous host cell membranes, so to the human body Ebola 'looks like another human cell'. Tricky, huh?

After entering your body, the GPs on the surface of the envelope interact with cellular walls of human tissue (eg. the nasopharyngeal lining, conjunctiva of your eye, blood, or even the vaginal endothelium) in a kind of molecular handshake that causes the virus particle and cell wall to merge. This allows the particle to insert its payload—an RNA genome that instantly begins to hijack your cell and transform its native proteins and cellular machinery into manufacturing copies of the Ebola genome and associated proteins. Among these are short glycoproteins, sometimes called secreted glycoproteins (sGP), which are manufactured in high numbers and then expelled from the cell into the bloodstream as 'counter measures' against the Immune System. This eventually results in a cytokine storm, where huge amounts of cytokine (SOS signals) are secreted, which eventually sends the immune system into fatal overdrive mode, but that is in the final stages of infection.

First the virus needs to amplify its numbers. The copied RNA genome also manufactures lots of longer glycoproteins (the kind that protrude from the virus's surface). These envelope glycoproteins are delivered to the host cell's membrane, which then form buds on its surface and push—picture lots of tiny finger-like projections which make the host cell look a bit like a koosh ball. Eventually, these buds force the host cell to burst, and lots and lots of new Ebola virus particles flood the area, like worms escaping a rotten apple. Each is wrapped in the host cell's 'uniform', and each carries an intact Ebola genome and is dotted on the exterior with freshly minted Ebola glycoproteins (courtesy of the now dead cell's genetic machinery).

Because Ebola loves connective tissues, the victim's collagen soon breaks down, removing all protection and structural support for internal organs. Blood clots form and the blood thickens throughout the body due to accumulation of dead, red blood cells. At the same time, endothelial cells that line arteries, capillaries, and veins, break down and die, and blood is released into the surrounding tissues.

Research into Ebola's pathogenesis is still sketchy due to the high risk to exposure and the small number of adequate BSL-4 labs in the US. As of today, we have just the CDC lab in Atlanta and the USAMRIID lab in Fort Detrick, Maryland with fully qualified BSL-4 facilities and personnel. However, as of Oct. 7, these two labs have partnered with 14 other US labs to form a diagnostic network for Ebola testing:

> The labs are selected members of the Laboratory Response Network, which was set up to allow rapid detection of a host of chemical or biological threats, including emerging infectious diseases.
>
> In August, the FDA issued an emergency use authorization for an Ebola test, using real-time polymerase chain reaction methods (RT-PCR), and the CDC asked several labs to gear up for testing using the assay, Mangal said.
>
> Aside from Texas, the labs now in the network include state facilities in Nebraska, Montana, Maryland, Florida, Minnesota, New York, Michigan, Virginia, North Carolina, Pennsylvania, and Washington, as well as local labs in New York City and Los Angeles County.[24]

While it's wonderful that our response system and lab testing capability is improving, each new trainee that handles Ebola must 'get it right' every time. One mistake can prove fatal.

But why is Ebola so deadly? If it's been around since 1976, why don't we have a cure? Can't the doctors and researchers on the frontlines provide research into the disease's means of infection? The answer is no. While it certainly makes sense for an 'on site' lab to be set up for viral research, the truth is that healthcare providers during an outbreak have little time to consider research or even record-keeping.

For instance, in a typical healthcare setting, every visit to a patient is followed by some type of notation. Nursing aids might chart vitals and input/output numbers (input is the amount of liquid consumed, while output tracks urine measurement—this is important when considering patient dehydration). Among other things, a nurse would chart patient's appearance, disposition, sleep habits, patient complaints or new symptoms, as well as medications administered and possible side effects observed or reported. Of course, all these used to be charted on a patient's paper record, but now most charting occurs on a computer.

Think about it. Imagine yourself on the frontline in an Ebola epidemic. How would you chart patient events and observations/actions in the field during an Ebola outbreak? Most of the time, these 'hospitals' are poorly constructed and understaffed. Sanitation is difficult. Ebola caregivers must wear some type of protective clothing—usually called 'Hazmat suits' (Hazmat means 'hazardous materials'). The exterior clothing must be disposable or sturdy, so that it can be hosed down with a bleach solution between wards. Face masks and three pairs of Latex gloves are recommended. Triple gloving can reduce dexterity, adding to the risk of needle sticks, and it also makes writing more difficult. However, if a glove 'fails' (has a fault from manufacturing or is compromised), then the caregiver can remove that glove and still have two for protection. At the very least, frontline personnel must double-glove.

As you are about to leave the ward, you want to make notes about your observations and actions. Do you dictate these? Do you make notes on paper? A computer? How would you sanitize a computer or handheld between wards? I have read of some units that are trying to keep records by carrying small handheld computers that are protected by plastic bags, and the bags are sprayed between wards/patients. Some workers have been trying to 'remember' vitals, observations, etc. and chart them after leaving the ward.

Ebola wards can be chaotic. Often, patients become combative or violent, as if the virus drives them into actions that increase the chance that viral particles can be disseminated into the area. Some patients rip off their clothing and try to escape—as if the virus drives them mad with fever and panic. In the final stages of an Ebola infection, most victims become lethargic, and caregivers might enter a ward only to find the floor (simple wood or concrete, and in the worst cases a dirt floor) covered with comatose or dead patients in various stages of undress.

I'll finish this chapter by saying that any person who volunteers to care for Ebola patients deserves our deep and abiding thanks and all our prayers. And while we're at it, let us pray that Ebola will burn out soon, and that it will not find purchase in western countries—but that genie may already be out of the bottle. In the next chapter, we will cover the story of Thomas Eric Duncan.

Come and see…

"Messieurs, c'est les microbes qui auront le dernier mot."

(Gentlemen, it is the microbes who will have the last word.)

—Louis Pasteur

Thomas Eric Duncan, the first person diagnosed with Ebola in the US and the first to die here of the dreaded disease, will I suspect if the Lord does not intervene to stop Ebola in its tracks, soon be listed as Patient Zero in the American Ebola epidemic. Before we begin his story, let's take a look at why so many Liberians now live outside their homeland.

Eric Duncan is one of thousands of Liberians who sought better lives outside their own, war-torn and poverty-stricken country. Liberia was originally founded as a refuge for slaves who left the United States to form an 'Eden' of their own in the early 1800s. Sierra Leone was established at about the same time by British freed slaves (1787). Though ostensibly independent, both countries remained tied to the political apron strings of their former nations. The World Wars (I and II) altered the geopolitics of Africa, particularly as Western nations required raw materials like native iron ore, timber, and rubber to produce war machines. Even today, the primary employer to many in Liberia is the Firestone Rubber Plantation.

In 1980, US-funded Master Sergeant Samuel Kanyon Doe led a *coup d'état* against incumbent president William R. Tolbert, murdering Tolbert and most of his cabinet. Doe became head of a military government, which favored US trade and helped to protect US assets in Africa.

In November, 1985, civil war broke out in Liberia, and racial tensions (hatred among tribal factions) led to an all-out tribal war by the end of the decade. As a result, the US withdrew funding to Liberia, but the wars continued, erupting into a second Civil War near the end of the twentieth century which continued until 2003. By then, *five percent* of Liberians were dead.

To secure new lives for themselves and their families, many Liberians immigrated to the west. In Dallas today, there are whole communities of Liberian immigrants.

According to an article published on October 5th at the New York Times:

> The murderous civil war that terrorized Liberia from 1989 to 2003 left at least 5 percent of the population dead, and

sent wave after wave of refugees to neighboring countries. To escape the ethnic and political turmoil, more than 700,000 fled from a nation that had barely two million residents when the conflict began.

Among them were Thomas Eric Duncan, the man who brought the Ebola virus to the United States last week, and Louise Troh, the woman he had come to Dallas to visit. After meeting in the early 1990s in a refugee encampment near the Ivory Coast border town of Danané, the two Liberians started a relationship and bore a son, several family members said.[25]

In Africa, Troh and Duncan's relationship hit rocky ground, and the couple broke up, and Louise Troh left Liberia to follow another man to Dallas, Texas in America. However, that new relationship ended, leaving Troh on her own to raise a family. Eventually, she and Duncan began to communicate again, and Eric Duncan decided to follow Troh to America and make her his wife.

[...]"They had had a falling out, and had patched things up," said the Rev. George Mason, Ms. Troh's pastor at Wilshire Baptist Church, "and he had come here with the intention to marry and start a new life together. Obviously, what happened has thrown a wrinkle into that."[26]

The human tragedy of Eric Duncan's life has played out in headlines across the globe. His story is similar to our fictional 'Mary', and one wonders how many contacts in Duncan's 'tree' may soon develop symptoms. To better understand the 'tree', let's examine Duncan's timeline:

- Thomas Eric Duncan (goes by Eric) quit his job on September 4th.
- 4 days *before* he departs Monrovia, Duncan helps his landlord transport Marthalene Williams, a pregnant woman to the hospital (she is the landlord's 19-year old daughter). Williams had been convulsing and was known to be infected by Ebola.
- The hospital turns Williams away, and Duncan helps her father carry Williams back into her home, where she died three days later. (According to the New York Times,[27] a neighbor reported seeing Duncan carrying the girl by the legs).
- Sept. 19th, Duncan is screened at Monrovia's airport for elevated temperature (registering as 97.3 degrees), and fills out a form asking if he'd touched or been in contact with an Ebola patient. Duncan says 'no'.
- Duncan's route takes him first to Brussels, then to Washington Dulles, and finally to Dallas. (It is noteworthy that on Oct. 3, 2014, ABC News[28] reported two cases of 'suspected Ebola' in the Washington D.C. area. Since then, the CDC reports that both patients—currently in isolation—have been cleared of having the disease.)

- September 20th, Duncan arrives in Dallas and takes up residence with Ms. Troh, her 13-year-old son, and two adult men (one is Duncan's nephew Oliver Smallwood, and the other is a friend Jeffrey Cole).
- September 24th, Duncan begins to feel unwell.
- September 25th, Duncan goes to the hospital with Ms. Troh. He tells the nurse that he has been to Africa. He is sent home with painkillers and antibiotics.
- September 28th, Duncan returns to the hospital, driven there by his niece who is a nursing assistant. He vomits on the sidewalk outside their apartment building. Once at the ER, Duncan is admitted, where he remained in isolation until his death on Oct. 8th.
- Ms. Troh, and all those in her apartment are told to remain inside. Some reports claim the CDC gave them little to no food. Ms. Troh told reporters that as of three days after Duncan's admission to the hospital, no one had even bothered to collect his soiled linens from her home, and no one had come there to disinfect or clean.[29] When an NBC reporter asks CDC director Tom Frieden about this, he replies "The details of that you'd have to refer to the folks in Dallas. But this is, after all, the first time we've ever had a case of Ebola in the US and there are issues to make sure that when things are removed, that it is not going to be disposed of in any way that could potentially be a risk."[30]
- A Dallas judge does supervise the clean-up of Ms. Froh's

apartment which commences on October 3rd, five days after Duncan's admission. The judge (Dallas County Judge Clay Jenkins, who is coordinating the Ebola response) said he said he was 'not happy' that family members were still quarantined at the home. Jenkins said Mr. Duncan's sweat-soaked bedsheets remained inside but were bagged and in the bedroom with the door closed.[31] Time Magazine online said this on Oct. 6th: "[Judge Jenkins] made a point of visiting the home of the infected man **without protective clothing**, took the responsibility for **driving his quarantined family to their new home**, and has been doing what he can to coordinate the state and federal response, while keeping his voters calm."[32] [Emphasis added]

- Oct. 3rd, Ms. Troh and all those who lived with her at her apartment home are moved to a quarantine home in a gated-community provided a local physician (unnamed probably to protect his privacy and to help keep the location of the home secret).
- Oct. 6th, Youngor Jallah, who is listed as Duncan's 'stepdaughter' is cleared to return to work as a nursing assistant even though she was exposed to 'violent vomiting' when taking her stepfather to the hospital.[33]
- Oct. 8th, Duncan is pronounced dead. He died on the same day as the second Blood Moon.
- Oct. 9th, MedPage Today reports that a Dallas Sheriff's Deputy is being evaluated in isolation for possible Ebola virus infection—he has since tested negative for the virus.[34]

As of today, October 9th, there is little additional information, except for news of a homeless man who has been apprehended and detained in quarantine because this man rode in the same ambulance as Duncan. It is assumed that the man, named as Michael Liveley, had been exposed to the virus. Liveley had been living and connecting with people since riding in the ambulance. He was found after a massive manhunt on Sunday, Oct. 5th. Just how many people came into contact with Liveley is currently unknown.

Ebola in America. It is a chilling thought. Anyone who has read Stephen King's horror novel, *The Stand*, may be tempted to compare the lightning speed of transmission and death for the 'Captain Trips' virus to Ebola, but Ebola virus is (currently) more difficult to transmit. Generally, a potential victim must have contact with blood, semen, saliva, vomitus, or feces of the patient—or touch a surface that contains samples of infected fluids.

Example: an Ebola patient sneezes or coughs, and virus-laden sputum lands on a book that you touch. You touch your eye or your nose/mouth, and the virus gains entry into your system. Unless it is killed with bleach solution or UV light, the Ebola virus can remain viable and capable of infecting someone for several DAYS on a surface.

News reports, the CDC, and the WHO insist that Ebola can only be spread through bodily fluids. As mentioned earlier in this book, Ebola Reston was presumably spread via the ductwork of the containment facility housing the monkeys. At least three humans who claimed to have no first-hand contact with the macaques fell ill, but only with minor symptoms. Subsequent blood tests on these humans were positive for Ebola Reston antibodies.

In *The Hot Zone*, writer Richard Preston's account of the Ebola Reston event, we read about a debriefing involving the CDC and USAMRIID scientists who responded to the 'incident'. They are asked if Ebola is airborne:

> Nancy Jaax described the incident in which her two healthy monkeys had died of presumably airborne Ebola in the weeks after the bloody-glove incident in 1983. There was more evidence, and she described that, too. In 1986, Gene Johnson infected monkeys with Ebola and Marburg virus by letting them breathe it into their lungs, and she had been the pathologist for that experiment. All of the monkeys exposed to airborne virus had died except for one monkey, which managed to survive Marburg. The virus, therefore, could infect the lungs on contact. Furthermore the lethal airborne dose was rather small: as small as five hundred infectious particles. That many airborne particles of Ebola could easily hatch out of a single cell. A tiny amount of airborne Ebola could nuke a building full of people if it got into the air-conditioning system. The stuff could be worse than plutonium, *because it could replicate*.[35] (Emphasis added)

Researchers at USAMRIID also reported in 1995 that they suspected Ebola may be transmissible by air:

TRANSMISSION OF EBOLA VIRUS (ZAIRE STRAIN) TO UNINFECTED CONTROL MONKEYS IN A BIOCONTAINMENT LABORATORY

This USAMRIID report, based on the same experiment, is published in *Lancet*, Vol. 346, December 23/30, 1995, p. 1669-1671:

SUMMARY:
Secondary transmission of Ebola virus infection in humans is known to be caused by direct contact with infected patients or body fluids. We report transmission of Ebola virus (Zaire strain) to two of three control rhesus monkeys (*Macaca mulatta*) that did not have direct contact with experimentally inoculated monkeys held in the same room. **The two control monkeys died from Ebola virus infections at 10 and 11 days after the last experimentally inoculated monkey had died.** The most likely **route of infection of the control monkeys was aerosol, oral or conjunctival exposure to virus-laden droplets secreted or excreted from the experimentally inoculated monkeys.** These observations suggest approaches to the study of routes of transmission to and among humans.[36] (Emphasis added)

In addition, two experiments, published in 2013 in the National Institutes of Health library PubMed, reported that guinea pigs and macaque monkeys who are exposed only to an aerosolized Ebola virus (that is a fine mist of the virus that 'hangs in the air') not only became ill, but **100 percent of them died.**

Our fictional example of 'Mary' earlier in this book might help as we consider the means of contracting Ebola. As with Eric Duncan, Mary had numerous contacts before her admission to the hospital. She touched many lives, who in turn touched many others, branching out in to a communication tree. If Ebola were easily transmissible, then Ebola would soon overtake our hospitals and overwhelm our healthcare system, but so far it is not easily communicated through the air.

But what if that changed? Can Ebola mutate? Or worse, what if someone weaponized it? Perhaps that has already happened. More in the next chapter.

Come and see…

Weapons based on viruses fell under the letter N. Smallpox for instance was described in letters in clandestine communications as N1. Ebola was N2, Marburg N3, and Marchupo or Brazilian Hemorrhagic Fever as N4.

—Ken Alibek, *Biohazard*[37]

"I looked up and saw the clock had gone from green to red, I turned on the monitor. Sadly, they're all—." He paused and looked at Baby LaVon's eyes, wide and although still red-rimmed with tears, curious. "They're all D-E-A-D down there."

—Stephen King, *The Stand*

We live in a country that prides itself on being open, fair, and democratic. You and I were taught from the cradle to respect government officials and trust in the rule of law to guide us and take care of us. We were also taught that 'other countries', especially those with communist ideologies have little regard for their own citizenry, so it seldom surprises us to read of bioweapons programs in books like

"Biohazard" by whistleblower Ken Alibek (born Kanat Alibekov), who worked for the bioweapons division *Biopreparat* in Russia until 1992, when he and his family defected to the United States. Our country helped Alibek to escape, and he has served in an advisory capacity to our CIA and other espionage networks. He also serves as CEO and Chief Science Officer at AFG Biosolutions in Gaithersburg, Maryland.

Alibek has testified before Congress numerous times and has also taught at George Mason University. His tell-all book, *Biohazard*, reveals how Russia kept their bioweapons research 'quiet' despite having signed an accord to stop all bioweapons programs in 1972.

On hearing this, most Americans would nod and say, "Sure, those commies always lie." But what happens when you discover that your own, trusted government has lied to you?

For years, the Department of Defense denied ever testing biological weapons on American soil, but in 2002, the Pentagon released documents revealing a covert American bioweapons plan that did include testing on American lands and American people:

> The summaries of more than two dozen tests show that biological and chemical tests were much more widespread than the military has acknowledged previously.
>
> The Pentagon released records earlier this year showing that **chemical and biological agents had been sprayed on ships at sea.** The military reimbursed ranchers and agreed to stop open-air nerve agent testing at its main chemical weap-

ons center **in the Utah desert after about 6,400 sheep died** when **nerve gas drifted away from the test range.**

Earlier this year, the Defense Department acknowledged for the first time that **some of the 1960s tests used real chemical and biological weapons, not just benign stand-ins.**[38] (Emphasis added)

Currently, the United States is still a signatory to and obligated by the Biological Weapons Ban Treaty of 1972. According to that document, only two countries in the world have permission to keep samples of BSL-4 biological entities—kept solely for defensive research in case of a pandemic. Those two countries are Russia and the United States. Since 1972, the geopolitical world has changed dramatically, but the ban is still in effect. In fact, the most recent review conference for the Bioweapons Ban Commission stated that:

"...**under all circumstances the use of bacteriological (biological) and toxin weapons is effectively prohibited by the Convention**"[39] (Emphasis added)

Yet, according to Alibek's book *Biohazard*, the Russian program not only experimented with Ebola after 1972, they managed to create sufficient quantities to deem it a possible weapon:

Our scientists had found it more difficult to cultivate Ebola than Marburg—they were not able to reach the necessary

> concentration, but by the end of 1990—the long-term problem of cultivation had been solved, and we were **close to developing a new Ebola weapon.**[40] (Emphasis added)

During the worldwide Smallpox Eradication Program, *vaccinia* virus was used as the agent of inoculation. Vaccinia is a member of the poxvirus family (Family *Poxviridae*), and infection with this mild pox virus can and does provide protection against the much more virulent and easily transmitted virus, Smallpox. The Smallpox Eradication Program (SED) ran globally from 1967 through 1979, at which time the worldwide problem was hailed as solved, and Smallpox named as the first virus completely and forever removed from the planet.

However, Smallpox virus samples are still kept in both the US and in Russia as research material just in case it ever does return, and vaccinia virus is stockpiled for such an event. Now, let me tell you what you probably don't know about vaccinia. It is a virus that can be used in conjunction with other viruses to create a **recombinant entity**. Again, I will quote from Alibek's unsettling, tell-all book, "Biohazard". In 1996, he tells us, a team of Russian researchers published an article in:

> "...*Molecular Biology*, a journal published by the Russian Academy of Sciences. The scientists reported that they had found a space in the vaccinia genome where foreign genetic material could be inserted without affecting virulence. They claimed the purpose of this research was entirely peaceful—

to explore different properties of the vaccinia virus, but what medical reason could there be for preserving its virulence?

The Vector scientists had used a gene for beta-endorphin, a regulatory peptide in their experiments. Beta-endorphin, capable in large amounts of producing psychological and neurological disorders and of suppressing certain immunological reactions, was one of the ingredients of the Bonfire Program. It was synthesized by the Soviet Academy of Sciences.

In 1997, the same team reported in the publication, *Questions of Virology*, that **they had successfully inserted a gene for Ebola into the genome of vaccinia.** Once again, a benign scientific explanation was put forward: they said it was an important step toward creating an Ebola vaccine. But we always intended for vaccinia to be our surrogate for further Smallpox weapons research. There was no doubt in my mind that **Vector was following our original plan. One of our goals had been to study the feasibility of a Smallpox-Ebola weapon.**[41] (Emphasis added)

Just writing this statement sends chills down my spine; and it should do the same to you. Vaccinia was the 'practice virus' for recombination. The real goal for Vector was to insert Ebola into Smallpox, making a deadly combination that is easily transmitted. This was in 1997. Seventeen years ago. Who knows what successes the Vector Program may have had since then?

Alibek no longer works for Russia, so he is no longer privy to insider information, but we can extrapolate with a certain amount of accuracy. In those seventeen intervening years, the science of genetics has grown exponentially. If Russia had the capability of inserting Ebola genes into vaccinia (a poxvirus that is very similar to Smallpox) at that time, is it possible—even probable—that they have successfully inserted an Ebola gene into Smallpox? I would have to say that the unsettling answer is yes.

No, I cannot prove that a Smallpox-Ebola bioweapon exists, but logic tells me that it is likely. Very likely.

Is the West Africa outbreak caused by a bioweapon such as the one described by Alibek? I doubt it. So far, it does not appear to be more communicable than wild-type Ebola. However, if certain disreputable scientists or even terrorists choose to release a recombinant Ebola, then how would we, the common public, know? Most likely, the sudden shift to an airborne Ebola would be explained away as a mutated virus.

Consider what the World Health Organization has been saying. An article at CNN in September said this:

> Dr. James Le Duc, the director of the Galveston National Laboratory at the University of Texas, said the problem is that no one is keeping track of the mutations happening across West Africa, so no one really knows what the virus has become.
>
> One group of researchers looked at how Ebola changed over a short period of time in just one area in Sierra Leone early on in the outbreak, before it was spreading as fast as it is now. They found more than **300 genetic changes in the virus.**[42] (Emphasis added)

President Obama has noticed, and he's responded by sending troops to West Africa. Deutsche Welle reported this:

> Every virus mutates to adapt to its surroundings. Ebola is no exception.
>
> When President Barack Obama announced the deployment of US forces to West Africa he said in an interview on Sunday that immediate intervention was vital.
>
> Otherwise, said Obama, **Ebola could mutate, making it more easily transmittable, "and then it could be a serious danger to the United States."**
>
> **The longer a virus circulates, the more it changes its genetic material.**[43] (Emphasis added)

The above quoted article seems to imply that Obama's deployment of troops will somehow prevent the virus from mutating. Good luck with that one. A virus doesn't care how many troops stand in its path—if it chooses to make you its next victim, then it will infect you.

The military 'brass' aren't impressed with President Obama's strategy either. In fact, an article posted to the news website *World Net Daily* on October 5, 2014 quoted Lt. Gen. William "Jerry" Boykin as being directly opposed to the idea of sending troops to fight a virus:

> "This is a president who thinks like a community organizer and not like a commander-in-chief who takes his responsibility

> for his troops seriously," Boykin said. "At a time when our military has been at war for 13 years, suicide is at an all-time high, [post-traumatic stress disorder] is out of control and families are being destroyed as a result of 13 years of war, the last thing the president should be doing is sending people into West Africa to fight Ebola."[44]

Here's another thought to ponder: If the bulk of the US military is serving overseas in one theater or another (the Middle East, Eastern Europe, and now Africa), who will keep order here if Ebola or another pathogen were to create a pandemic and subsequent panic within our borders? We have seen a tiny forerunner of such a scenario playing out in Dallas. A military guard was stationed at the apartment home of Ms. Troh (Thomas Eric Duncan's girlfriend) to keep her and the others who lived there from running away. The article continued to quote Boykin's amazement that the US appears to be heading into this fray alone:

> "What about the other African nations? What about a coalition? Where is the U.N.?" Boykin asked. "There should be a U.N. coalition to try to stem the tide of this Ebola (epidemic). This isn't a U.S. military operation, and it should not be a U.S. military-led operation."[45]

Now, she and those she loves are in quarantine—under guard. However, oddly enough, others who were exposed to Duncan—one who even drove him to the hospital after he had begun exhibiting symptoms, a nursing assistant named Youngor Jallah[46]—have been given the all clear by the CDC. I ask you: would you feel comfortable having Ms. Jallah care for your loved one? Ebola has an incubation period of up to *twenty-one days*. I assume that the CDC has examined a blood sample from Ms. Jallah, but there is always a possibility that the virus had not yet multiplied to sufficient levels for her blood test to be reliable.

I pray the CDC is not making mistakes at this critical time. I hope that our government health agencies will always err on the side of caution. Honestly, I'm shocked that authorities did not put everyone associated with Duncan into quarantine the moment he was diagnosed. In the CDC's own lab, if someone who works with deadly pathogens like Ebola is involved in an 'incident', that person is placed in the slammer (quarantine) for a month.

I've often told the listeners to our online internet talk program, *PID Radio*, that I have a tinfoil hat that I keep in my pocket to wear as needed. I will put on my hat for a moment now and speculate—understand, this is pure speculation on my part, but it is speculation based on what I know so far.

Consider then the possibility that a weaponized version of Ebola does exist in a lab somewhere—perhaps in Russia, perhaps seized by a terrorist—or perhaps in our own country (after all, some scientists who work with pathogens might rightly conclude that Russia has been concocting biological weapons, so why shouldn't we?—we certainly don't want a bioweapons gap!). If such a lab-created pathogen did exist, then an outbreak of wild-type Ebola in West Africa is opportunistic—a chance for a field 'test'. And what was it Rahm Emanuel said?

> You never let a serious crisis go to waste. And what I mean by that it's an opportunity to do things you think you could not do before.[47]

No, I'm *not* blaming Emanuel for Ebola (although he is a lousy Chicago mayor), but I mention this because Emanuel's belief system has heavily informed and underwritten that of our current president, Barack Obama.

And since I am now wearing my tinfoil hat for speculation purposes, let me mention that the Affordable Care Act, which Nancy Pelosi famously said had to be passed so they could see what was in it, may prove the very control device implemented if Ebola were to take root in the United States. (If you doubt that Pelosi actually said this, see a video of her comment to Congress at YouTube[48]).

This is the scenario I foresee (my tinfoil hat firmly in place):

1. An initial infection in a major US city leads to an outbreak of small pockets of infections within the area. This is the 'Mary' scenario stated earlier in this book that an Ebola patient who does not develop symptoms quickly can touch many lives who in turn touch many others.
2. As Ebola leads every newscast, we begin to hear experts warning that Ebola may mutate into a deadlier, more efficient pathogen.
3. One or two companies seek permission to field test their vaccines on one or two of these patients (still in the 'trial' stages)

4. These vaccines appear to grant the recipient mild to moderate immunity from Ebola, and the Department of Health and Human Services (HHS) orders the company to produce as many doses as possible.
5. The company or companies agree, but the process is painstaking and slow, so they can only produce the vaccine in small batches—perhaps a million doses a month. The government orders them to begin making as much as possible and the world clamors to place orders.
6. The Affordable Healthcare Act uses these doses to create rings of immunization around the outbreak region(s). This follows the Smallpox Eradication Program Protocol. Those inside the outbreak zone are quarantined—no one goes in, no one comes out. A ring of immunization at a distance from the outbreak area is created—like a firebreak.
7. The plan requires that all those within the outbreak area must be quarantined. Enforcement would fall on either local police or on the military. The people inside this area would be cut off from the world.
8. Within the quarantine zone, the virus would run rampant. Water, fuel, and food shortages would only play to the advantage of the pathogen, further weakening immune systems. People would panic, crime would increase, and the world would watch from a distance while a pocket of humanity, a region of culture and civilization and western advantages slowly dies.

9. Outside the hot region, the immunization ring may or may not work, but as more vaccine is produced, eventually everyone in the US, and possibly in the world, would be required to receive an inoculation. It would be a requirement placed upon all healthcare providers, and since more people are now insured (thanks to the Affordable Healthcare Act), more Americans are subject to legal requirements. If you do not comply, your coverage will be terminated.

In truth, the only leverage required to begin vaccine enforcement consists of 1) a pathogen so terrifying and a scenario so horrible that people will panic even though no disease is yet in their area, and 2) the presence of hope in a bottle, a vaccine.

Want to know more? Come and see…

And he opened the bottomless pit; and there arose a smoke out of the pit, as the smoke of a great furnace; and the sun and the air were darkened by reason of the smoke of the pit.

—REVELATION 9:2

A Bush administration intelligence review has concluded that four nations—including Iraq and North Korea—possess covert stocks of the smallpox pathogen, according to two officials who received classified briefings. Records and operations manuals captured this year in Afghanistan and elsewhere, they said, also disclosed that Osama bin Laden devoted money and personnel to pursue smallpox, among other biological weapons.

—BARTON GELLMAN, WASHINGTON POST WRITER

Smoke is a mysterious entity. Nearly, invisible, it can take on the shape of its container, fill every space, every crack, every crevice; it can crawl up a wall, spiral into a column, or boil into a ball of noxious particles; and it can kill. In the book of Revelation, the Apostle John compares the escape of demons from their prison, the Abyss, to 'smoke from a furnace'. Smallpox can travel the same way. Like smoke—from a furnace. It can and does enter wherever, whenever it chooses, and it can strike with an invisible hand, infecting and as such killing a new victim—someone who is simply in the wrong place at the wrong time.

Richard Preston is sometimes accused of over-dramatizing his non-fiction accounts of Ebola and Smallpox, but the facts contained within his beautifully written pages cannot be denied—else he'd have been sued again and again. In *The Demon in the Freezer*, Preston tells how smoke is used to determine how Smallpox managed to infect persons on a completely different floor from an isolated male patient in a small German hospital.

Smallpox is a true, airborne disease. It can live outside a human host for up to twenty-four hours, which means anyone breathing that air or touching a surface contaminated by that air runs the risk of infection. In December 1969, a man whom Preston calls 'Peter Los' entered St. Walberga hospital in Meschede, Germany with an unknown illness. The man had recently traveled to Pakistan, where a Smallpox epidemic was underway. Los had no idea that he had brought a hitchhiker along with him from Kurachi.

As the doctors and nurses worked with Los, they had no reason initially to suspect Smallpox, but assumed he must have typhoid:

> ...Monday and Tuesday passed. Every now and then a nun would come in and collect his bedpan. His throat was red, and he had a cough, which was getting worse. The back of his throat developed a raw feeling, and he sketched restlessly. At night he may have suffered from dreadful, hallucinatory

dreams. The inflamed area in his throat was no bigger than a postage stamp, but in a biological sense it was hotter than the surface of the sun. **Particles of smallpox virus were streaming out of oozy spots in the back of his mouth, and were mixing with his saliva. When he spoke or coughed, microscopic infective droplets mixed with smallpox particles were being released, forming an invisible cloud that floated in the air around him.** Viruses are the smallest forms of life. They are parasites that multiply inside the cells of their hosts, and they cannot multiply anywhere else. A virus is not strictly alive, but it is certainly not dead. It is described as a life form. **There was a cloud of amplified virus hanging in his room, and it was moving through the hospital.** On Wednesday, January 14th, Peter's face and forearms began to turn red.

The red areas spread into blotches across Peter Los's face and arms, and within hours, the blotches broke out into seas of tiny pimples They were sharp feeling, not itchy, and by nightfall they covered his face, arms, hands, and feet. Pimples were rising out of the soles of his feet and on the palms of hands, and they were coming up in his scalp, and in his mouth, too. During the night, the pimples developed tiny, blistery heads, and the heads continued to grow larger. They were rising all over his body at the same speed, like a field of barley sprouting after rain. They were beginning to hurt dreadfully, and they were enlarging into boils. They had a waxy, hard look, and they seemed unripe. His fever soared abruptly and began to rage. The rubbing of pajamas on his skin felt like a roasting fire. He was acutely conscious and very, very scared. The doctors didn't know what was wrong with him.

By dawn on Thursday, January 15th, his body had become a mass of blisters. They were everywhere, all over, even on his private parts, but they were clustered most thickly on his face and extremities. This is known as the centrifugal rash of smallpox. It looks as if some force at the center of the body is driving the rash out toward the face, hands, and feet. The inside of his mouth and ear canals and sinuses had pustulated, and the lining of the rectum may have pustulated, as it will do in severe cases. Yet his mind was clear. When he coughed or tried to move, it felt as if his skin was pulling off his body, that it would split or rupture. The blisters were hard and dry, and they didn't leak. They were like ball bearings embedded in the skin, with a soft, velvety feel on the surface. Each blister had a dimple in the center. They were pressurized with an opalescent pus.

The blisters began to touch one another, and finally they merged into confluent sheets that covered his body, like a cobblestone street. The skin was torn away from its underlayers across much of his body. The blisters on his face combined into a bubbled mass filled with fluid, until the skin of his face essentially detached from its underlayers and became a bag surrounding the tissues of his head. His tongue, gums, and hard palate were knobbed with pustules, yet his mouth was dry, and he could barely swallow. The virus had stripped the

> skin off his body, both inside and out, and the pain would have seemed almost beyond the capacity of human nature to endure it.[49] (Emphasis added)

Smallpox, like Ebola, is a tiny virus, but it packs a supernatural punch. While Los was quarantined in his hospital room, an unlucky visitor to Los's floor merely opened the door at the end of the hallway to ask a nurse for directions. He then closed the door and descended to a different floor. That man contracted Smallpox. On the floor above Los, a nun contracted Smallpox. To determine how these people contracted the illness despite the quarantine, the hospital hired an expert in smoke.

The smoke man arrived at St. Walberga, and he set his smoke machine inside Peter Los's room, opening the window a crack just the way Los had done without the nurse's noticing (though he admitted it later). The doctor in charge was shocked as he and the smoke man followed the smoke out the room and down the hallway:

> ...and a cloud of black smoke poured out of the nozzle and headed for Los's door and billowed down the hallway of the isolation ward. Paul Wehrle ran along with it. The smoke went through the cracked-open door and poured into the lobby, and from there it boiled up the stairs to the second floor and then went to the third floor. As it came out of the stairwell, it drifted along the upper hallways. It got through the closed doors of the cloistered hallway on the third floor, and it sprinkled a number of sick nuns with black dust.

> "The patients got more of a treatment than they'd bargained on when they went to the hospital," Wehrle said to me. "They were individually sooted with high-grade soot." [...] Meanwhile, Richter and Posch had gone outdoors and were standing on the lawn. Wehrle heard them shouting, and he opened the window and looked out.
>
> The smoke was seeping outdoors under the raised casement and flowing in a thin, fanlike sheet up the walls of the hospital. Wehrle ran around and began opening the upper windows just a crack. To his amazement, the smoke came into the upper rooms from outside, having **crept up the walls.**[50] (Emphasis added)

Smoke rose up and infected the other floors of the hospital, finding its path through cracks and crevices, and even creeping up the walls. Smallpox virus is like smoke particles, carried aloft and outside, and like smoke, it can creep up a wall.

A virus is not a living entity, but it has a purpose. That devious purpose is to find a host, assume command of as many cells as possible, and then reproduce itself millions of times so that it can then infect other hosts. A virus is like a machine. A machine whose sole *raison d'etre* is to multiply.

So, what if Russia did make a Smallpox-Ebola hybrid? We don't know that this chimera actually exists, though Ken Alibek would have us believe it does. I will say this. Smallpox on its own, were the virus ever released into a dense population, would decimate that area and worse. Smallpox disease also has a hemorrhagic component, sometimes causing the victim's skin to detach due to pooling blood beneath the dermis (this is sometimes called 'Black Pox' because the skin turns black). Some researchers have even theorized that it was 'Black Pox' that actually ravaged Europe in the 14th and 15th centuries, not *Yersinia pestis* (also known as 'Bubonic Plague').

A few paragraphs back, I mentioned a news article from October 6th about the first Ebola transmission outside of Africa. A nurse in Spain has taken ill and tested positive for the disease. This morning, additional reporting has revealed that the precautions taken by healthcare providers may not have been according to WHO protocols. The Guardian reports:

> Staff at the hospital told El País that the protective suits they were given did not meet World Health Organisation (WHO) standards, which specify that suits must be impermeable and include breathing apparatus. Staff also pointed to latex gloves secured with adhesive tape as an example of how the suits were not impermeable and noted that they did not have their own breathing equipment.[51]

So what does a health care worker (HCW) wear when caring for an Ebola patient? DuPont, a company that makes PPEs (Personal Protective Equipment) lists the following:

> All persons entering the patient room should wear at least:[52]
> - Gloves
> - Gown (fluid resistant or impermeable)

- Eye protection (goggles or face shield)
- Facemask
- Additional PPE might be required in certain situations (e.g., copious amounts of blood, other body fluids, vomit, or feces present in the environment), including but not limited to:
  - Double gloving
  - Disposable shoe covers
  - Leg coverings

The bottom line is that you want to keep any virus from hitching a ride into your personal life. Did you cut your nails this morning? Is it possible that you nicked the surrounding tissue—even the tiniest bit? That's a port of entry for a virion. Cut yourself shaving? Point of entry. Unshielded eyes? Point of entry. Uncovered nose? Point of entry for a patient who coughs right into your face (droplets of sputum have been shown to contain Ebola virions).

The double-gloving may seem like overkill (pardon the pun), but researchers in BSL-4 labs often triple glove—this is a precaution against glove failure, which happens far more often than one might imagine.

Face respirators are not required in Ebola epidemics, since Ebola is not currently efficient at infecting humans via lung tissue, but if I were working with Ebola patients, I think I'd want one—just in case. Not only must a HCW consider his or her own life, but also the lives of the many others he or she might infect as a consequence.

The Blood Serpent is biting, and it will probably continue to bite for months to come. How do we prepare? And what do *Thanatos*, the Death rider and his hellish companion *Hades* have planned? What have they been given permission to do?

Come and see...

And I looked, and behold, a pale horse! And its rider's name was Death, and Hades followed him. And they were given authority over a fourth of the earth, to kill with sword and with famine and with pestilence and by wild beasts of the earth.

—REVELATION 6:8 (ESV)

He's got a ticket to ride, and he don't care.

—PARAPHRASE OF LYRIC FROM "TICKET TO RIDE", THE BEATLES

The English word blood appears 447 times in the King James Version of the Bible. The Hebrew that is generally translated as 'blood' is dam (pronounced 'dawm'), and it bears an interesting meaning, both obvious and subtle. Dam actually means 'silent'. Isn't that interesting? The first time it appears in the Old Testament is in Genesis 4:10, when God asks Cain where Able has gone. God says to Cain, "What hast thou done? The voice of thy brother's blood crieth out to me from the ground." Remember, 'dam' actually means silence, and yet this silent witness is crying out to God!

I find that comforting. Blood is the life, God tells us (Deut. 12:23). And blood can testify to things that happen in our lives. It has healing power—for it is the blood of Christ that covers all our sins. Mankind is forbidden to shed blood of another man, and he is forbidden to drink blood. Blood is a lifespring that originates with God—it is a lifeline. And, 'silent' blood can speak volumes to Him.

On October 8th, many of us gazed heavenward as the second Blood Moon of the 2014/2015 Blood Moon Tetrad[53] rose in the nighttime sky. Mark Biltz is the author who first noted how four lunar eclipses would occur during this time period, a sabbatical year for Israel, and that each eclipse would occur during a holy season.

God has provided the stars and the heavens for beauty, but also for discerning the signs and seasons (Genesis 1:14). The Hebrew word for 'seasons' is *mow'ed*, which implies a divine appointment. By the end of October 2014, we will have seen two solar eclipses and two lunar eclipses. Here is the schedule:

April 15, 2014—Total Lunar Eclipse
April 29, 2014—Annular Solar Eclipse
October 8, 2014—Total Lunar Eclipse
October 23, 2014—Partial Solar Eclipse

In 2015, we'll see these eclipses:

March 20, 2015—Total Solar Eclipse
April 4, 2015—Total Lunar Eclipse
September 13, 2015—Partial Solar Eclipse

September 28, 2015—Total Lunar Eclipse

In Matthew 24:29, when Jesus was speaking to the disciples, He said this:

> Immediately after the tribulation of those days shall the sun be darkened, and the moon shall not give her light, and the stars shall fall from heaven, and the powers of the heavens shall be shaken.

Revelation 6:12–15 echoes this prediction and repeats this 'sign':

> And I beheld when he had opened the sixth seal, and, lo, there was a great earthquake; and the sun became black as sackcloth of hair, and the moon became as blood; And the stars of heaven fell unto the earth, even as a fig tree casteth her untimely figs, when she is shaken of a mighty wind. And the heaven departed as a scroll when it is rolled together; and every mountain and island were moved out of their places. And the kings of the earth, and the great men, and the rich men, and the chief captains, and the mighty men, and every bondman, and every free man, hid themselves in the dens and in the rocks of the mountains;

When a lunar eclipse occurs, it turns red making it appear as if the moon has 'turned to blood'. When a total solar eclipse happens, it turns 'black as sackcloth'. If these events are meant to shake us out of our comfort zones, then we need to pay attention to the lunar and solar events such as the Blood Moon tetrad, an event that is extremely rare when falling on high holy days.

A total solar eclipse, according to Jewish teachings, means judgment for 'the nations'. The total solar eclipse earlier this year was on March 20th. Take heed to what was reported regarding the timeline for the West African Ebola outbreak:

> In Paris, the samples were taken to the Institut Pasteur. But the institute reported a technical problem at the lab and had to move the samples to another facility, 250 miles away, in Lyon, where technicians were put on alert and told to wait up. Once the samples arrived, the technicians worked into the night. By a little after two A.M. on **March 20**, they had the first results: what they were looking at was a filovirus, meaning it couldn't be Lassa. **Later that day, at seven P.M., the worst was confirmed: the samples were positive for Ebola.**[54] (Emphasis added)

The official beginning to the 2014 Ebola outbreak in West Africa is March 20th, the same day as the solar eclipse. Does that mean that Revelation 6 is being played out before our eyes? I'd have to say that it is a possibility, but it is more likely that we are receiving a warning—a foreshadowing of what is to come. As mentioned earlier about the 'Braxton Hicks' contractions that come weeks before the actual birth, these events may be a warning to those paying attention to the signs.

But whether the ride has already commenced or is about to occur, the solemn truth is this: Thanatos has been given permission to ride, and Hades follows in his wake to eagerly swallow up the dead brought to him by his minion, Death.

The African problem could soon become the problem of the entire world. I believe that these regional outbreaks and the near-miss in Reston are alarm bells meant for all of us to hear. The birth pangs, the signs of the end of the age that Christ Himself warned us about are getting closer together and much harder. The final seven years of the World as We Know It are fast approaching, and it is up to you and me—up to all believers—to cry out that now is the accepted time of salvation.

The riders draw near. The White Horse and rider want to conquer the world. Politically, we see Russia gobbling up the former Soviet Union states such as Crimea and now Ukraine. ISIS, now called simply the Islamic State, has declared its intention to take Jordan, Saudi Arabia, Syria, Lebanon, and all of the Levant, including Israel.

The Israelis struggle to maintain their borders and their sanity. China claims seas near Japan, and Japan claims waters near Korea. Despite proclamations by the UN listing Africa as top of their list for aid, this third world collection of disparate nations suffers crippling poverty and serves more as a major shopping mall for first world nations who covet Africa's vast wealth in diamonds, oil, natural gas, precious metals, and recently discovered rare earth minerals used in electronics and laser manufacture.

As Christ promised birth pangs that would increase in intensity over time, our world is suffering with major 'Braxton Hicks' contractions! Currently, we are hearing pre-echoes of what is to come, but one day soon, they WILL RIDE.

The white rider will steal peace and sovereignty, the red rider will bring war, the black rider will ruin economies and bind all under a fascist regime led by AntiChrist, and Thanatos, who rides the sickly green horse will wield his weapons and prey upon the now physically weakened populations, dragging them into the jaws of Hades.

Over the past six months or so, thousands of illegal immigrants—some children, some much older, some gang members—have overrun our borders and now sit in detainment facilities while our government decides how best to handle the invasion. No one knows what tiny traps, what *therion* these immigrants may be carrying. Recent news reports indicate that at least some carry tuberculosis and measles. Many have coughs and fevers, but not all have been quarantined. Border agents and doctors complain that it is impossible to adequately evaluate all who come through the centers before each group is loaded into a bus or plane and sent on to another holding area.

The Lord loves every one of the children and adults who have crossed into the United States. He does not care that they are here illegally or legally. God allows human government to remain in place—for now. Honestly, if I were living in Central America or Mexico and thought my children would have a better life elsewhere, I would try with all my might to get them there. I am not blaming the immigrants. I blame the politicians for the mess in Texas and other border states right now. And it is becoming clear that many if not all of the immigrants now in the system will eventually arrive in our hometowns. These children may soon learn in our local schools and live in neighborhoods all across the United States, hoping to integrate into our society.

Will they be healthy? Will they bring exotic illnesses to which you and I have no immunity? How will this affect our economy? Our society? Our laws? We are already a surveillance nation, and an influx of unknown quantities and potential gang affiliates will only increase the tension between the watchers and the watched.

The white rider seeks to conquer the United States through political trickery. It is beginning to look, prophetically at least, like our country is doomed to fail. If we capitulate and become part of a larger entity such as a North American Union or a Pan America, then the world as we know it will change radically—in a very short period of time.

The Red Rider will bring a balkanized form of tribal warfare to our land. Civil war and rebellion will collapse what is left of our economy, leaving most Americans without jobs, without homes, without food. We already see hints of this in the angry responses to the Obama administration's legislative and executive actions, but the Red Rider has no affiliation with any political party; both sides of the aisle seem hellbent on collapsing the dollar and the American way of life. The Black Rider's yoke and oppression of corporatism will weigh even heavier upon the shoulders of all who see the Riders.

And then comes the final rider, Thanatos, astride his sickly steed bringing plague and pestilence. His *rhomphaia* will slice through humanity, destroying whole populations. Power will be given to

Thanatos to kill twenty-five percent, *one-fourth*, of the Earth's population! One quarter of all mankind!

Yes, this scenario is bleak and it may sound hopeless, but you and I know that there is always Hope, and His name is Christ Jesus! His sinless life, sacrifice, and resurrection have removed Death's sting! John, the disciple who loved Jesus as a brother and to whom Christ entrusted his mother at Calvary, has given us the prophetic foreknowledge of what is to come; the vision given to Him by Jesus Christ and addressed to all the members of His Body, the true Church. The slain Lamb enters the heavenly court to prove that HE IS THE TRUE CONQUEROR! It is His nail-scarred hands that break each seal. He allows the troubles the Earth will soon face, but all these woes are intended to drive humanity to Him in supplication and prayer and penitence. It is a refining fire meant to purify.

I know that some of what I've discussed here is conjecture, but I have no doubt about this: Christ Jesus *is* returning soon for His Bride. Soon thereafter, the four horses, white, red, black, and pale will begin to ride throughout the Earth leaving devastation and death in their wake. We who believe, who call Christ our Savior and King must pray, preach, teach, and warn those we love, those who will listen—yes, all those whom God has prepared by His Spirit with ears to hear. Churches no longer teach the book of Revelation because the enemy does not want us to be forewarned. Therefore, we few who watch and yearn for that glorious appearing must bridge that gap. We must begin to fast and pray, to meditate upon His word, to seek His wisdom, so that we can teach the Truth while there is yet time.

But we must also prepare. Food stores, printed Bibles, commentaries, perhaps digital versions on thumb drives or SD cards. Medical supplies, water, blankets. Prepare as if your home may one day become an *ad hoc* hospital. How can you do this?

Come and see…

"Do you find viruses beautiful?"
"Oh, yeah," he said softly. "Isn't it true that if you stare into the eyes of a cobra, the fear has another side to it? The fear is lessened as you begin to see the essence of the beauty. Looking at Ebola under an electron microscope is like looking at a gorgeously wrought ice castle. The thing is so cold. So totally pure."

—Richard Preston, *The Hot Zone*

For centuries Ebola had lurked in the jungles of central Africa. Its emergence into human populations required the special assistance of humanity's greatest vices: greed, corruption, arrogance, tyranny, and callousness.

—Laurie Garrett, *The Coming Plague*

Death is often seen as an impersonal, uncaring entity. I'd say this assessment is probably correct. Death has a job, an assignment, one might say, to perform according to God's instructions. However, you and I needn't fear 'The Reaper', because those who place their faith in the blood of Christ have His protection. Nothing reaches us that is not permitted by God.

However, that does not mean that we shouldn't prepare. During the coming months—if the world experiences a pandemic (a worldwide outbreak of Ebola or a variant)—then we can prepare in advance to be ready to help and share the gospel. We've been warned. We stand on the wall and watch the horizon for God's warning. We look to the skies for His signs, and it's clear that something Evil is on the way.

So, how do we prepare our homes and our hearts for what is to come? Before I give you a list of items and their uses, let me ask you to imagine your daily routines. I've been considering mine for weeks now, picturing how virus transmission could occur within my family.

How often do you clean your surfaces with bleach? How often do you wash your hands—do you scrub your nails every time? Do you turn off the unclean faucet with your clean hands? Do you touch an unclean doorknob after washing? When you wash your dishes, are they sterile? Do you hand-wash or use a dishwasher? How clean are your floors? Your linens? Do you open your door to strangers and converse with them?

At work, do you drink from a common water fountain? Do you share a coffee pot? Are those cleaned regularly? When you shop at the grocery store, do you touch fresh vegetables that have possibly been touched by another? Do you clean the shopping cart handle? Do you touch the checkout conveyor belt? Do you touch the payment screen?

Beginning to see my point? Each moment throughout our day, we face the possibility of infection. An Ebola outbreak makes every interaction with people and surfaces a chance to contract death itself. So, let's discuss how we can prepare for the worst—and *doing so soon* is advised, since waiting until an epidemic is declared may be too late to find or afford many of these items.

Let's begin with your home.

Start now to add these to your home's storeroom:

1. Non-perishable food stocks, water, matches, and a battery operated radio (a hand-crank radio is a good choice, and one that receives short wave is best). Even if your area is not affected by an outbreak, commerce and transport will likely be compromised. Most stores keep only a three-day supply in their stores at any time, so a breakdown in deliveries will mean empty shelves in a hurry, so lay in a supply that can last for at least a month or more if you have the space and the funds.
2. Plastic sheeting, available in most home and garden supply stores such as Lowe's or Home Depot. Get the black plastic rolls, not the plant barrier cloth that is permeable. You want to provide a 100 percent barrier to virus.
3. Face masks, Hazmat suits, plastic gloves. You can buy these online and the cost is relatively small. If you cannot afford the Hazmat suits (which are Tyvek or similar material and disposable), then plan on burning or bleaching anything you wear during homecare for an infected patient. Nursing scrubs in any color can usually

be bleached. Buy several boxes of unsanitary plastic (Latex) gloves, because you will want to wear at least two pairs at a time and dispose of them.

4. Construction grade plastic bags for disposal of discarded gloves, etc. and to dispose of any soiled linens or clothing.
5. A large heavy plastic trash can with a tight fitting lid that will be lined with the above bags. You don't HAVE to buy a trash can, but it will make it easier to keep the black bags open without the need for someone to hold them. And you can keep a lid on anything inside until it is burned, laundered, or buried.
6. Duct tape—lots of it. Buy a case. You will use this to tape up plastic 'doors' and to secure your disposable gloves.
7. Plastic mattress protectors. These will help keep bodily fluids from infecting your mattress. And while we're discussing this, you should choose a room to serve as an isolation chamber. Select a bed (a rollaway or inflatable would work perfectly) that will work for the patient. Keep mattress protectors on (double them, so you remove one and keep the second in place, then add a new one so you always have two). Buy inexpensive white sheets that can be bleached or that you can afford to burn. Ebola patients have temperature variances (chills early on, but they become very 'hot' later and tend to throw off covers). In advanced stages, the patient will

have almost constant diarrhea, so you will be glad for the cheap linens and plastic cover.

8. Bedpan and a urinal.
9. Have a cleanup kit with lots of old rags, bleach, and an insecticide or paint sprayer that you can fill with bleach solution and use to 'spray down' anyone who has been in contact with the patient.
10. Vinegar and perhaps even a sunlight lamp. UV light kills Ebola as does 3% acetic acid. A bleach solution can be prepared using bleach that is at least 2.5% strong, though 5% strong is best. The bleach solution degrades after twenty-four hours, so plan to make fresh batches daily as needed. You will need this solution sprayer to decontaminate any surfaces, items, or even people who have come into contact with the patient.

You'll need to provide a 'sick room' which can be separated from the rest of the house. A storage room, perhaps even a garage, provided that it can be kept warm or cool, depending on the season. Scripture tells us that we are to love others as we love ourselves, so try to provide comfort for the sick with a comfortable room. Our garage is insulated, and it could be isolated easily from the rest of the home, so that is where I'd set up our 'sick room' if needed. Consider a space where your patient's remains can be safely removed (if he/she dies, and remember Ebola has a 50-90% fatality rate) without risk of contamination to other spaces. A garage door permits easy access to this.

You will also need:

1. Bandages, adhesive tape, disinfectant such as alcohol for soaking combs, scissors, and any item that comes into contact with the patient such as a thermometer (by the way, plan to dispose of this thermometer once your patient no longer needs it, and DO NOT use any thermometer that you use on an Ebola patient with anyone else in your home).
2. Skin disinfectants such as iodine, Betadine, peroxide, 70% Isopropyl alcohol, hexachlorophene, and eye irrigation liquids (saline, such as that used by contact lens wearers—you can buy this in large bottles, keep watch on the expiration dates). Antibacterial soap should also be on-hand.
3. Tylenol or the generic (acetaminophen). This is the ONLY SAFE pain reliever, because all other NSAIDS (including aspirin and ibuprofen) are blood thinners. The Ebola patient has a virus that is already causing hemorrhaging, so the last thing he/she needs is more blood thinning!
4. Disposable face masks and eye protection can be purchased online. Wear one every time you interact with your patient. Clean your goggles with alcohol.
5. Also, make sure you keep all these supplies in the space that you plan to use as an isolation space.

Setting up your isolation space requires forming a buffer zone between your home and the 'sick room'. If you use the garage, then you can exit the garage into the yard or driveway and dispose of all 'contaminated materials' there. It is also a convenient place to spray yourself or have someone spray you with a bleach solution if needed. Eye protection is also recommended. You can purchase lab goggles inexpensively online. If you do not wear eye protection and are splashed with vomitus, blood, or other bodily fluids, then you should irrigate your eyes with clean saline and have a helper spray the rest of you with the solution—or you could take a hot shower with antibacterial soap.

The buffer zone can be created indoors by taping black plastic sheeting in the doorway (you would still close the door). Then tape more black plastic a few feet away from this space to form the 'buffer zone'. For instance, we have a bedroom which opens to a hallway. We could use that bedroom as the 'sick room' and the hallway as the 'buffer'. Just remember that your buffer zone should not be a common space that your family still uses. In our home, there is also a bathroom in the hallway, so that could be a cleanup space, but it would be off limits to anyone not infected or caring for the patient.

Finally, you will need to dispose of soiled items, and there may come a time if a national pandemic breaks out, that people will have to shelter and be quarantined at home, which also may require that we each dispose of those who have passed away. West Africa is now burning most bodies, but this may not be possible here. In that case, some areas may have organized patrols looking for bodies—yes, I know this is not what you want to think about, but you must think about it now—not later. Funerals may take place in our backyards, and we must use the horror of this thought to propel us to action!

Death is not the end. Death is not the final winner. Scripture tells us this:

> He will swallow up death in victory; and the Lord GOD will wipe away tears from off all faces; and the rebuke of his people shall he take away from off all the earth: for the LORD hath spoken it.
>
> —Isaiah 25:8

> So when this corruptible shall have put on incorruption, and this mortal shall have put on immortality, then shall be brought to pass the saying that is written, Death is swallowed up in victory. O death, where is thy sting? O grave, where is thy victory?
>
> —1 Corinthians 15:54-55

I believe that Christ will return any time—in the blink of an eye—and all who know Him as Savior will meet Him in the clouds. Some believe in a pre-trib catching away, but we who watch for His appearing may find our society breaking down in advance of that glorious day! Be ready, dear friends, to minister through His Word, through your open hearts, open hands, open homes.

The harvest truly is plenteous, but the laborers are few. The White Rider *will* ride forth. The Red Rider *will* follow. The Rider in Black *will* descend upon those oppressed by corporatism's yoke, and the fourth Rider, Thanatos and his companion Hades *will* ride forth to bring plague and pestilence and death, but you and I can run in advance of their rides, before they reach that lost person, that lost child, that lost neighbor. We can cut them off at the pass through God's word. Read it, preach it, give it away while we still have time.

Even so, come, Lord Jesus.

Recommended Reading:

The Hot Zone by Richard Preston

The Coming Plague by Laurie Garett

Survive the Coming Storm: Ebola Crisis by Ray Gano

Biohazard: The Chilling True Story of the Largest Covert Biological Weapons Program in the World—Told from Inside by the Man Who Ran It by Ken Alibek

Vaccine A by Gary Matsumoto

The Demon in the Freezer by Richard Preston

"Case Definition for Ebola Disease" a CDC factsheet for healthcare workers found online at http://www.cdc.gov/vhf/ebola/hcp/case-definition.html

"Sequence for Putting On Personal Protection Equipment (PPE)" a CDC Poster available online at http://www.cdc.gov/vhf/ebola/pdf/ppe-poster.pdf

"Checklist for Patients Being Evaluated for Ebola Virus Disease (EVD) in the United States" a CDC publication available at http://www.cdc.gov/vhf/ebola/pdf/checklist-patients-evaluated-us-evd.pdf

"Infection Prevention and Control Recommendations for Hospitalized Patients with Known or Suspected Ebola Virus Disease in the US" a CDC factsheet available at http://www.cdc.gov/vhf/ebola/hcp/infection-prevention-and-control-recommendations.html

"Specific Laws Concerning the Control of Communicable Diseases" a CDC publication, available at http://www.cdc.gov/quarantine/specificlawsregulations.html

"Legal Authorities for Isolation and Quarantine" a CDC publication available online at http://www.cdc.gov/quarantine/aboutlawsregulationsquarantineisolation.html

Notes

1. For more on 'the Nachash', see Michael S. Heiser's excellent article via http://michaelsheiser.com/TheNakedBible/2010/02/the-absence-of-satan-in-the-old-testament/ (accessed Oct. 8, 2014)
2. Excerpt from "The Viking Portable Greek Historians" found at Wikipedia, link: http://en.wikipedia.org/wiki/Plague_of_Athens (accessed Oct. 2, 2014)
3. http://www.poemhunter.com/poem/book-vi-part-04-the-plague-athens/ (accessed Oct. 2, 2014)
4. David W. Lowe has a simple yet profound article on the linguistic source of the word and the concept of 'The Rapture' online at http://www.earthquakeresurrection.com/excerpts/04.rapiemurandharpazo.pdf (accessed Oct. 8, 2014)
5. Daniel's 70 Week prophecy is found in Daniel 9. For a quick summary of this prophecy, see Chuck Missler's article "The Precision of Prophecy" at http://www.khouse.org/articles/2004/552/ (accessed Oct. 3, 2014)
6. To learn more about the Blood Moon Tetrad and why it is important to prophecy scholars, see Mark Biltz's bestseller, "Blood Moons", available here: http://www.amazon.com/Blood-Moons-Decoding-Imminent-Heavenly/dp/1936488116/ref=sr_1_1?ie=UTF8&qid=1412342849&sr=8-1&keywords=blood+moon+biltz (Amazon link was accessed Oct. 3, 2014)

7. Braxton Hicks contractions are named for the John Braxton Hicks, the English doctor who first discovered them. For more on this phenomenon of pregnancy, see 'What Are Braxton Hicks Contractions?' at the online site 'baby center' via http://www.babycenter.com/0_braxton-hicks-contractions_156.bc (accessed Oct. 3, 2014)
8. For more on the Rwandan Genocide and RTLMC Radio, see the Wikipedia entry on the station at http://en.wikipedia.org/wiki/Radio_T%C3%A9l%C3%A9vision_Libre_des_Mille_Collines (accessed Oct. 3, 2014)
9. Rosenberg, Jennifer, "A Short History of the Rwandan Genocide" published at about.com, available online at http://history1900s.about.com/od/rwandangenocide/a/Rwanda-Genocide.htm (accessed Oct. 3, 2014)
10. Much of the Yambuku story is derived from reading the two books, *The Coming Plague* by Laurie Garrett (see below) and *The Hot Zone* by Richard Preston.
11. Airborne Ebola has not been proven to the satisfaction of most virologists, but a study by Twenhafel, et al, published at National Institutes of Health Pub-Med website, titled "Experimental Aerosolized Guinea-Pig Adapted Zaire Ebolavirus (Variant Maygna) Causes Lethal Pneumonia in Guinea Pigs" indicates that aerosolized Ebola transmission is a very real possibility. To read the Asbtract, see: http://www.ncbi.nlm.nih.gov/pubmed/24829285 (accessed Oct. 8, 2014)
12. Ibid.

13. For more on aerosolized transmission, see also: Twenhafel, et. al, "Pathology of Experimental aerosol Zaire ebolavirus infection in Rhesus macaques", published by the National Institutes of Health PubMed online library, available via http://www.ncbi.nlm.nih.gov/pubmed/23262834 (accessed Oct. 8, 2014)
14. Roberts, John, "Ebola Outbreak 'Out of Control' Says CDC Director", originally published on Sept. 2, 2014 at Fox News, available online via http://www.foxnews.com/health/2014/09/02/ebola-outbreak-out-control-says-cdc-director/ (accessed Oct. 8, 2014)
15. Levine, Marianne, "WHO can't deal with Ebola outbreak, health official warns", originally published on July 17, 2014, at Los Angeles Times website, available via http://www.latimes.com/world/africa/la-fg-who-ebola-20140718-story.html (accessed Oct. 8, 2014)
16. ZMapp is an experimental drug that has shown promise in combating Ebola Hemorrhagic Fever. The drug was 'fast-tracked' by the WHO and approved for administration to a select few patients. Dr. Kent Brantley, a physician with Samaritan's Purse, who contracted Ebola in West Africa was flown back to the US and given ZMapp. Dr. Brantley recovered and has since been voluntarily donating his blood sera to other victims. For more on ZMapp, see the Wikipedia entry at http://en.wikipedia.org/wiki/ZMapp (accessed Oct. 8, 2014)
17. Garrett, Laurie, *The Coming Plague: Newly Emerging Diseases in a World out of Balance*, Farrer, Straus, and Giroux, New York, 1968,

page 148. This author used the Kindle edition, sold by Macmillan through Amazon books. Link: http://www.amazon.com/The-Coming-Plague-Emerging-Diseases-ebook/dp/B005FGR6RO

18. Alibek, Ken, "Biohazard.." see above—Location 2192.
19. Russell, Kevin et al, "The Global Emerging Infection Surveillance and Response System (GEIS), a US Government Tool for Improved Global Biosurveillance: A Review 2009" Published online at http://www.ncbi.nlm.nih.gov/pmc/articles/PMC3092412/ (accessed Oct. 9, 2014)
20. CDC Factsheet for Hospitals: Interim Guidance for Environmental Infection Control in Hospitals for Ebola Virus, originally published Aug. 1, 2014, last updated Oct. 3, 2014, CDC Website http://www.cdc.gov/vhf/ebola/hcp/environmental-infection-control-in-hospitals.html (accessed Oct. 6, 2014)
21. By Staff, BBC News report on the Hajj, "Saudi Arabia Plays Down Ebola Concern for Hajj Pilgrimage" found online at http://www.bbc.com/news/world-middle-east-29461229 (accessed Oct. 4, 2014)
22. Writer unlisted, "Exclusive: Thousands from Ebola Nations Allowed to Enter US Without Additional Screening", published at Breitbart News on October 2, 2014, available online via: http://www.breitbart.com/Breitbart-Texas/2014/10/2/Exclusive-Thousands-from-Ebola-Nations-Allowed-to-Enter-US-Without-Additional-Screening (accessed Oct. 4, 2014)
23. Ibid.
24. Smith, Michael, "Ebola: Lab Net Ready to Test", originally

published on Oct. 7, 2014 at MedPage Today, available online via http://www.medpagetoday.com/InfectiousDisease/Ebola/47980 (accessed Oct. 8, 2014)

25. Sack, Kevin, "Ebola Victim's Journey from Liberian War to Fight for Life in U.S.", published October 5, 2014 online at The New York Times, available via http://www.nytimes.com/2014/10/06/us/ebola-victim-went-from-liberian-war-to-a-fight-for-life.html?partner=rss&emc=rss&smid=tw-nytimes&_r=0 (accessed Oct.5, 2014)
26. Ibid.
27. Ibid.
28. Mohney, Gillian, "Ebola Scare Hit Washington, Latest of About 100 Alerts to CDC", published October 3, 2014 online at ABC News, available via http://abcnews.go.com/Health/ebola-patient-tested-washington/story?id=25948325 (accessed Oct. 3, 2014).
29. Associated Press, "Armed Guards Blocking Family Exposed to Ebola from Leaving Dallas Home", published at New York Post online via http://nypost.com/2014/10/03/armed-guards-blocking-family-exposed-to-ebola-from-leaving-dallas-home/ (accessed Oct. 6, 2014)
30. Ibid.
31. Boyle, Louise & Bates, Dan, "Hazmat Team Arrives at Ebola Victim's Apartment after Five Days...", published at the Daily Mail online via http://www.dailymail.co.uk/news/article-2779036/Hazmat-team-arrives-Ebola-victim-s-apartment-FIVE-DAYS-later.html (accessed Oct. 3, 2014)

32. Nicks, Denver, "This Texas Judge Is Fighting Fear and Ebola in Dallas", originally published on Oct. 6th at Time online, available via http://time.com/3474650/ebola-dallas-judge-jenkins/ (accessed Oct. 9, 2014)
33. Collins, Laura, "Dallas Ebola Victim's Stepdaughter...Given All Clear to Return to Work as Nursing Assistant", originally published on Oct. 6, 2014 at DailyMail, available via http://www.dailymail.co.uk/news/article-2782694/Ebola-victim-s-stepdaughter-took-hospital-vomiting-wildly-given-clear-return-work-nursing-assitant.html (accessed on Oct. 6, 2014)
34. Smith, Michael, "Ebola: Second Texas Man Assessed", originally published on Oct. 8, 2014 at MedPageToday, available online via http://www.medpagetoday.com/InfectiousDisease/Ebola/48000?xid=nl_mpt_DHE_2014-10-09&utm_content=&utm_medium=email&utm_campaign=DailyHeadlines&utm_source=ST&eun=g712026d0r&userid=712026&email=sharonkgilbert%40gmail.com&mu_id=5874041&utm_term=Daily%20B (accessed Oct. 9, 2014)
35. Preston, Richard, "The Hot Zone" pp. 221-222 Anchor Books Reprint Kindle Edition (Mar 4 2012), purchased at Amazon via link: http://www.amazon.com/Hot-Zone-Richard-Preston-ebook/dp/B007DCU4IQ/ref=sr_1_1?ie=UTF8&qid=1412429374&sr=8-1&keywords=the+hot+zone (accessed Oct. 4, 2014)
36. USAMRIID Report "Transmission of Ebola Virus (Zaire Strain) to Uninfected Control Monkeys in a Biocontainment Laboratory, originally published at Lancet, Vol. 346, December 23/30, 1995,

pp 1669-1671. Found online via http://www.ncbi.nlm.nih.gov/pubmed/8551825 (accessed Oct. 4, 2014)

37. Alibek, Ken, with Handelman, Stephen, "Biohazard: The Chilling True Story of the Largest Biological Weapons Program in the World, Told From Inside By the Man Who Ran It", Kindle Edition, location 376. Available at Amazon via http://www.amazon.com/Biohazard-Chilling-Largest-Biological-World--Told/dp/0385334966/ref=sr_1_2?ie=UTF8&qid=1412616788&sr=8-2&keywords=biohazard (link last accessed on Oct. 6, 2014). Note that this Kindle version does not have real page numbers, but only uses location numbers.
38. Associated Press, "Bioweapons Tested in US in 1960s" available online via http://usatoday30.usatoday.com/news/nation/2002-10-09-bioweapons-tested-sixties_x.htm (accessed Oct. 6, 2014)
39. Wikipedia entry on "Biological Weapons Convention", available online via http://en.wikipedia.org/wiki/Biological_Weapons_Convention (accessed Oct. 6, 2014)
40. Alibek, Ken, "Biohazard.." see above - Location 2192.
41. Ibid. Location 4166.
42. Cohen, Elizabeth, "Ebola in the air? A nightmare that could happen", originally published on Oct. 2, 2014 by CNN online via http://www.cnn.com/2014/09/12/health/ebola-airborne/ (accessed Oct. 6, 2014)
43. Staff, "Unstoppable: Is Ebola mutating with unknown consequences before our eyes?", originally published on September 10, 2014 at Deutsche Welle website, available via http://

www.dw.de/unstoppable-is-ebola-mutating-with-unknown-consequences-before-our-eyes/a-17912329 (accessed Oct. 6, 2014)

44. Maloof, F. Michael, "General Blasts Obama's Order of Troops to Fight Ebola", originally published on October 5, 2014 at World Net Daily, available via http://www.wnd.com/2014/10/generals-blast-obamas-order-of-troops-to-fight-ebola/ (accessed Oct. 6, 2014)
45. Ibid.
46. Collins, Laura, "Dallas Ebola victim's stepdaughter—who took him to the Dallas hospital as he was vomiting wildly—is given all clear to return to work as nursing assistant", originally published on October 6, 2014 at Daily Mail Online, available via http://www.dailymail.co.uk/news/article-2782694/Ebola-victim-s-stepdaughter-took-hospital-vomiting-wildly-given-clear-return-work-nursing-assitant.html (accessed Oct. 6, 2014)
47. Yandle, Bruce, "Rahm's Rule of Crisis Management: A Footnote to the Theory of Regulation", originally published on Feb. 11, 2013 at The Freeman, available via http://fee.org/the_freeman/detail/rahms-rule-of-crisis-management-a-footnote-to-the-theory-of-regulation#axzz2KizozPJ4 (accessed Oct. 6, 2014)
48. Nancy Pelosi said, "But we have to pass the bill, so that you can, uh, find out what's in it; away from the fog of all this controversy." YouTube video of this statement can be found at https://www.youtube.com/watch?v=hV-05TLiiLU (accessed Oct. 6, 2014)
49. Preston, Richard *The Demon in the Freezer,* pp. 33-35, Random House, published 2002, Kindle version used, purchased at http://www.amazon.com/Demon-Freezer-

True-Story-ebook/dp/B000QCSANQ/ref=sr_1_1?s=digital-text&ie=UTF8&qid=1412639067&sr=1-1&keywords=demon+in+the+freezer

50. Preston, Richard, *The Demon in the Freezer*, pp.56-57, Kindle Version used, see above for more.
51. Kassam, Ashifa, "Spanish Nurse's Ebola Infection Blamed on Substandard Equipment", posted on Oct. 7, 2014 at The Guardian online, available via http://www.theguardian.com/world/2014/oct/07/ebola-crisis-substandard-equipment-nurse-positive-spain (accessed Oct. 7, 2014)
52. Dupont Technical Bulletin, "Protective Clothing for Ebola (EVD)", published 2014, available online as pdf via http://www.dupont.com/content/dam/assets/products-and-services/personal-protective-equipment/assets/DPP14_20240_Ebola_Tech_Bulletin_%2091114b.pdf (accessed Oct. 7, 2014)
53. For more on the blood moon tetrad, see the YouTube Video of J.R. Church's and Gary Stearman's interview with author Mark Biltz: https://www.youtube.com/watch?v=32ZkCygd8bE (accessed Oct. 7, 2014) Biltz's book is available at Amazon via http://www.amazon.com/Blood-Moons-Decoding-Imminent-Heavenly/dp/1936488116/ref=sr_1_1?s=books&ie=UTF8&qid=1412689792&sr=1-1&keywords=mark+biltz (accessed Oct. , 2014)
54. Stern, Jeffrey E., "Hell in the Hot Zone", originally published in October 14 issue of Vanity Fair, available online via http://www.vanityfair.com/politics/2014/10/ebola-virus-epidemic-containment (accessed Oct. 7, 2014)

CPSIA information can be obtained at www.ICGtesting.com
Printed in the USA
LVOW10s2309051114

411786LV00001BB/1/P